*A Look
at the
Economy of Nature
and the
Ecology of Man*

The Forest and the Sea

Marston Bates

*With a New Introduction
by Loren Eiseley*

TIME Reading Program Special Edition
TIME INCORPORATED · NEW YORK

COVER DESIGN *Matt Greene*

This book is dedicated to the students of Zoology 235 at the University of Michigan. They have contributed greatly to my education; have given me continuing friendships; and have provided an assurance that, as far as the next generation is concerned, the future is in good hands.

Contents

The Forest and the Sea

Editors' Preface

One day in the 1940s, as a biologist named Marston Bates worked at the top of a ladder that reached some 80 feet into the arboreal canopy of the South American rain forest, a surprising thought flashed through his mind. Although his treetop post was fairly hot and several hundred miles inland, Bates found himself thinking of the sea. This vision was brought on not by a longing for a cooler place but by a fresh scientific notion—a suddenly sharpened awareness of the parallel behavior of marine and woodland creatures. More than a decade after that moment of insight, Bates's idea developed into *The Forest and the Sea,* a deceptively entertaining book which may someday alter man's view of the world around him.

Across the pages of this book march endless parades of strange and wonderful creatures: mosquitoes that rise and sink in the depths of the forest like plankton in the sea; fish that travel over land; barracuda that lurk in the black waters of night; army ants more "devastatingly fierce" than any other imaginable predator. A honeybee sees things in invisible ultraviolet light, and a monkey fresh from its native habitat learns to steal spectacles from strangers' noses and make an omelet on the ceiling. But *The Forest and the Sea* is far more than merely a kaleidoscopic adventure or a clever comparison of fowldom with fishdom: it is a mature and creative expression of humanity's place in nature.

At the time Bates entertained the germinal idea which led to the title of this book, he was well on his way to becoming one of history's most knowledgeable experts on the distinctive manners and morals of one of the world's most unlovable inhabitants, the mosquito. Indeed, he was in the treetops when the idea struck him because he was investigating a high-flying species of this insect. He had earlier spent five years studying malarial mosquitoes in Albania and Egypt, and was now in the midst of an eight-year stay in a remote spot of Colombia to investigate the role of the pest in jungle yellow fever. The consequence, in 1949, was Bates's first book, *The Natural History of Mosquitoes,* a thoroughgoing scientific exploration of mosquito behavior. It was immediately acclaimed as a fundamental contribution to science. Among other things, it dramatized by implication how man's hopes of controlling his environment could be dashed if he ignored the actual relationships between different species. (For example, Bates demonstrated why the Rockefeller Foundation's early hope of eradicating jungle yellow fever simply by suppressing it in population centers was a forlorn dream: in the treetops, in the absence of man, certain mosquitoes kept the virus alive by spreading it among monkeys.)

Addressed to other scientists, that first volume merely hinted at the importance of man's involvement in the behavior of animals and the realities of nature. *The Forest and the Sea,* spawned in the same mosquito-laden current and destined for missionary duty in the same cause of natural history, is quite different. In one sinuous flash through the literary stream it offers substance for the seeker of exotic curiosities, a lesson for the molecular biologist, philosophy for the intellectual and the wisdom of an old-fashioned naturalist for the lover of literature. The book examines the plant and animal worlds, their relationship to one another, and man's relationship to both of them.

Perhaps the broad canvas Bates employs to depict his planetary view of nature is a direct product of his lifelong dual interest in science and writing. He published his first scholarly article, "A Geometrid Larva on Grapefruit," while in high school, then as a college undergraduate tried vainly to place all manner of poetry, stories and articles with popular magazines. He has now produced nine books, including the tome on mosquitoes and *The Forest and the Sea*. He also headed a group of scientists who produced a new high school biology text (which instead of beginning with the traditional—and dull—body cell, starts out with a familiar scene from nature, a rabbit sitting beneath a raspberry bush). Altogether, more than 100,000 copies of his books have been sold, and most of the titles are still in print. *The Forest and the Sea,* originally published in 1960, has been translated into Japanese, Portuguese, Spanish, Arabic, Polish and French.

Bates has scant patience with scientists who complain that laymen do not understand them. The root of their problem, he suggests, is an uncontrollable habit of spewing out an inky cloud of Greek terms behind which any meaning can hide like an imaginary octopus. "Science remains a rather mysterious affair," he has remarked in *The Nature of Natural History,* "cultivated by special priesthood, guarded by an unintelligible jargon." In *The Forest and the Sea* he expands the point.

> We are liable to fool ourselves into thinking we have produced a new thought, when we have only produced a new word. . . . Biocenosis leads easily to biomes and biochores, to ecosystems, ecotones and seres. These are all lovely words, but they don't really say anything new. The trouble is that the word-coiner, sinking blissfully into his addiction, gradually loses all communication with the outer world.

Bates is equally concerned about the fragmentation of knowledge. Biologists today are sharply split into two camps: the smaller number of them, including Bates, concentrate on the relationships among organisms or, to give the subject its learned name, ecology; the rest have given up such time-honored pursuits in favor of a heady quest for truths at the level of the molecule. To most molecular biologists a mosquito is primarily something to be repelled by a chemical on the skin. To the ecologist, a mosquito is, among other things, an influence on the world he lives in and on its other inhabitants—including mankind. As Bates says in this book:

> Man has not escaped from the biosphere. He has got into a new, unprecedented kind of relationship with the biosphere; and his success in maintaining this may well depend not only on his understanding of himself, but on his understanding of this world in which he lives. . . . It looks as though, as a part of nature, we have become a disease of nature—perhaps a fatal disease. . . . I am not advocating a return to the neolithic. . . . But long run efficiency would seem to require certain compromises with nature.

As anthropologist Loren Eiseley points out in his eloquent new introduction to this special edition, Bates has seen the implications of humanity's relationship to nature with a clarity that even Darwin failed to achieve. "Marston Bates has literally put man in his place in a way nobody before him has done," says virologist Telford H. Work of the U.S. Public Health Service Communicable Disease Center. "He has put together what biologists know in a way that every intelligent person in the world can understand. You simply cannot assess the effect this has had on key people in many walks of life."

As part of this effect, more and more people have been led by Bates to an awareness of the dangerous extent to which the human race is upsetting nature. "Surely," he says in *The Forest and the Sea,* "men who can manufacture a moon can learn to stop killing each other; men who can control infectious disease can learn to breed more thoughtfully than guinea pigs; men who can measure the universe can learn to act wisely in handling the materials of the universe."

—THE EDITORS OF TIME

Introduction

Modern technological man increasingly sets up unexpected reverberations in his universe. More and more, a kind of dissonance is communicated from the human world of invention to the world of nature. The interlocked web of life begins to vibrate with an ever-mounting rapidity. In the most scientific age in history we are losing the ability to marvel at any but our own creations. "The great palace of nature" over which 18th Century writers exclaimed so vividly has been supplanted, in many minds, by a superficial adulation of our own short-lived cleverness.

Today the essayist in the realm of natural history frequently is ignored as an artist even though his books are widely read. Some of this indifference may arise through a feeling on the part of literary critics that they are inadequate to deal with the subject matter. Still, these same critics do not hesitate to probe the symbols inherent in the novelist's art. All too frequently, it is merely an esthetic fashion that turns them away from the man of literature whose field is science.

Marston Bates, author of *The Forest and the Sea,* is such a man. A professional biologist of distinction, he can also make a mosquito the legitimate subject of literature. Not the least of his talents is the possession of a Baconian ability both to dilate and to contract the eye—something that great Elizabethan scholar held to be the true key to scientific observation. Bates belongs with the scientific travelers of the 19th Century, but he has returned to the jungle with the advantage of an additional century of information to draw upon; he deserves to be read with his earlier namesake, Henry Walter Bates, and with Alfred Russel Wallace.

Consider, for example, his treatment of the Amazon, which he justly calls not merely a river, but rather a kind of vast, sprawling inland sea. Here, because of the huge outpouring of fresh into oceanic waters, a curious chemical doorway has opened, one of those accidental chinks in nature through which living creatures sometimes slip, with a minimum of adaptation, from one environmental zone to another—in this case from salt water to fresh. The Amazon waters contain, as a consequence, a fresh-water dolphin and even a fresh-water sting ray—one of those strange sea bats which we do not normally expect to find nestling in such an unlikely place as a sand bar in the Andean uplands.

Bates also discusses those fascinating relatives of the pineapple known as bromeliads—plants which grow upon trees high in the forest and which catch and hold quarts of water, creating by their numbers another sort of dispersed water system far above and between the Amazonian tributaries. These plants promote the development of an insular water fauna as diverse as the evolutionary oddities which arise in the course of time on oceanic islands. The plants are really innumerable little laboratory boxes in which a kind of microevolution is going on. Or, again, Bates may dwell on comparable events taking place in the hidden waters within bamboo stems—worlds where the mosquito which frequents such places must fire its eggs with bull's-eye accuracy through tiny holes in order to reach the life-sustaining water, and where the adult, after its transformation, has to exercise similar ingenuity in order to escape its bamboo prison.

Besides contracting his eye and mind to the microscopic dimensions necessary to make out the life history of one species of mosquito out of hundreds, Bates can dilate his mind to encompass entire geological eras, enabling him to consider, among other things, the rise of man and the follies which that restless biped threatens to inflict upon his world.

I have said that there is something in Bates's approach to the awe-inspiring diversity of life on our planet, particularly in the tropics, which compares favorably with the best scientific writing of the voyager-naturalists. They were learning about the web of life, but they had no means of realizing how quickly scientific technology and the rapidity of human increase were destined to alter the wild places of the earth.

With one or two exceptions, the 19th Century scientists saw man's evolutionary rise as a natural event impossible to control consciously. They seem not to have visualized the possibility that man himself might direct the life about him on a major scale. Not even Darwin seems to have realized that man was on the point of escaping the forces which dominate the rest of life and was about to alter the face of the planet beyond recall.

Today indissolvable detergents return to us in our drinking water, we drop radioactive wastes into the sea, oil from ships befouls the wings of sea birds. Pesticides find their way into our own tissues. The wet lands that nourish wild waterfowl give way to drained and "improved" land as population multiplies, while radioactive strontium from atomic explosions pollutes our milk and enters our bones. Superhighways increasingly thrust the green world under cement. As Bates remarks, yellow-fever mosquitoes descended from the treetops when the woodcutters came. And always, everywhere, as the human swarm proliferates, sanitary engineering grows more difficult, water more precious, and by the same token, unpolluted water more difficult to come by.

In recent times, however, a persistent few have raised their voices, not alone against the reckless consumption of irreplaceable resources, but also against the whole philosophy that man can stand totally apart from nature. He cannot array himself against the old green world that made him and escape unscathed from her embrace. Man, too, is part of nature; he, too, draws his energy from the sunlight on the leaf; he, too, feels comfort in walking under the quiet of great trees at evening.

It is one of the terrors of our urbanized civilization that within it arises the man totally alienated from nature. Food comes from shelves, animals are strange things in cages before which one makes faces in the zoo. There is no surcease of noise. Daylong and nightlong, subways screech, trucks rumble, people shout. Outside is the green world, a world of little sunlit particles which, in every meadow leaf or in the wide pastures of the sea, are turning sunlight into life. Outside is the quiet, the quiet of an old rock in the sun. It is for these things that the minority has begun to express concern, to say, "Man has an ethic toward man, however badly he misuses it at times. He knows good from evil in human relationships, but toward the dust from which he came, the sunlight in his eyes, the breath that warms his lungs, he has no ethic."

Man has lived within nature until now, and taken her for granted. He has lived with nature like an unquestioning child. This is no longer enough. Man must now face the prospect of destroying nature and, in turn, being destroyed, or he must learn to protect and cherish for himself and unborn generations this beautiful planet with all its strange lifeways from which he has been granted the privilege of emerging. Marston Bates happens to be one of those farseeing people who glimpse the hope of a new pact between ourselves and mother earth. Behind the calm lucidity of *The Forest and the Sea* is the passion of one who loves life in a myriad of forms beyond his own.

There is an episode in the book which is especially revelatory of this aspect of Bates the man. Finding a strange reef creature one day on a Pacific atoll, he asked a marine biologist what it was. "It's one of the improbable animals," his colleague replied humorously. This set Bates to thinking. Perhaps he should write an essay upon improbable creatures. But then he began to look with a new perspective upon everything. He decided the essay was impossible to produce because all the living things in the world, including those in his own back yard, were improbable. "I had to go to Micronesia," he adds ruefully, "to get this particular view of my back yard."

I can tell one more anecdote from my own store of memories. At a scientific conference full of pontificating scholars of eminence, it came Bates's turn to speak. "I don't think we know all about this matter," he said, referring to some point in genetics which involved human evolution. "I think the geneticists know a very great deal, but not that much. I think it is better if we do not fill up our ignorance with words." Doubtless I am paraphrasing from memory, but the moment has stayed with me many years. Bates was the only scientist at that gathering willing and able to say simply that for all our hard-won knowledge we were still in the midst of a great mystery.

The reader will not be long in sensing a similar deep and unpretentious sincerity in this book, for, as is true of all good writers, the man and his book are indivisible. Moreover, there is in this union something that partakes of literature and therefore of an expressed humanity which passes outward beyond the boundary of the laboratory into that world of wind and water which modern man will neglect only at his peril.

The Forest and the Sea contains the kind of elementary knowledge which a good wizard would strive to impart to novices setting foot on a new planet, whose mysterious forces they were inclined to ignore. Bates is warning us gently about the intricate chain of living matter of which we constitute a part, and he is saying, as did Francis Bacon over 300 years ago: "Force maketh Nature more violent in the Returne."

—LOREN EISELEY

The Forest and the Sea

1. The Study of Life

[Biology] is the least self-centered, the least narcissistic of the sciences—the one that, by taking us out of ourselves, leads us to re-establish a link with nature and to shake ourselves free from our spiritual isolation.

—JEAN ROSTAND, in *Can Man be Modified?*

People often come to me with some strange animal they have found.

"What is it?" they ask.

Frequently I can't say—sometimes I get a despairing feeling of never knowing the answers to questions people ask. But at least I knew where to look it up; or I know someone who is an expert on that kind of animal, so I can relay the question to him. And once in a while I know the answer.

"Oh," I say brightly, "that is a swallowtail butterfly, *Papilio cresphontes*."

It is curious how happy people are to have a name for something, for an animal or plant, even though they know nothing about it beyond the name. I wonder whether there isn't some lingering element of word magic here, some feeling that knowing the name gives you power over the thing named—the sort of feeling that leads members of some savage tribes to conceal their personal names from all except their intimates. An enemy, learning their name, might be able to use this power for some evil purpose.

But other questions follow, "Where does it live?" and "What does it do?"

I explain that it is a tropical butterfly, common in Florida, which sometimes gets quite far north in the United States. The caterpillar lives on plants of the orange family, and north of Florida the butterfly is usually associated with prickly ash, which is a relative of the orange.

Almost inevitably there will come another question, "What good is it?"

I have never learned how to deal with this question. I am left appalled by the point of view that makes it possible. I don't know where to start explaining the world of nature that the biologist sees, in which "What good is it?" becomes meaningless. The question is left over from the Middle Ages; from a small, cozy universe in which everything had a purpose in relation to man. The question comes down from the days before Copernicus' theories removed the earth from the center of the solar system, before Newton provided a mechanism for the movements of the stars, before Hutton discovered the immensity of past time, before Darwin's ideas put man into perspective with the rest of the living world.

Faced with astronomical space and geological time, faced with the immense diversity of living forms, how can one ask of one particular kind of butterfly, "What good is it?"

Often my reaction is to ask in turn, "What good are you?"

Science has put man in his place; one among the millions of kinds of living things crawling around on the surface of a minor planet circling a trivial star. We can't really face the implications of this, and perhaps it is just as well—though I think humility is in general improving for the human character. A billion years into the past and a billion light-years into space remain abstractions that we can handle glibly, but hardly realize. We remain important, you and I and all mankind. But so is the butterfly—not because it is good for food or good for making medicine or bad because it eats our orange trees. It is important in itself, as a part of the economy of nature.

The question ought to be, not "What good is it?" but "What is its role in the economy of nature?" I like that phrase "the economy of nature," though there is a special word for the study of the interrelations of living things, *ecology*. Both words come from the Greek *oikos,* meaning household; both can have narrow and special meanings, but both can also be used broadly. Economics can be thought of as the ecology of man; ecology as the study of the economy of nature. This is one aspect of biology, one aspect of the study of life. It is thus also one aspect of science.

The word science covers a multitude of activities. We usually group these varied activities into three broad classes, which we call the physical sciences, the biological sciences and the social sciences. This is reasonable. The physical sciences deal with matter and forces in the natural world. The points of view, the methods, the objectives, change somewhat when we add the element of life. These change again when we turn to man and have to deal with the added element of culture, of behavior governed by accumulated tradition and transmitted through speech and symbol systems.

Ultimately it may be possible to explain man in biological terms, and life in physico-chemical terms—to reduce all the complexities of poems and wars and bird songs to mathematical equations. But we are a long way from this and we find it most convenient now to operate rather differently at these different levels. Curiously, the physicists, sure that the ultimate answers will be theirs, tend to be a little scornful of the muddling biologists; and the biologists, convinced that man is an animal, are dubious about the fancy studies of the social scientists. But this only shows that scientists are human beings, and science another human activity.

The division into physical, biological and social sciences looks logical enough, but it runs into all sorts of difficulties when we start to use it. What do we do with sciences like biochemistry and biophysics? And on the other hand, sciences like psychology and anthropology find themselves dealing with man as an animal as well as with man as a bearer of culture. Even with our present inadequate knowledge, it is clear that we are dealing with a continuum of natural events and that any division, however useful, is also artificial—reflecting the needs of the human mind rather than the realities of nature.

The difficulties—and the unrealities—increase when we turn to biology itself and to the problem of distinguishing among the different ways of studying life, which are the different kinds of biological sciences.

Sometimes I wonder whether there is any such thing as biology. The word was invented rather late—in 1809—and other words like botany, zoology, physiology, anatomy, have much longer histories and in general cover more coherent and unified subject matters. But while I doubt that biology has achieved a real existence yet, I am sure that botany, zoology, physiology and anatomy ought not to exist. I would like to see the words removed from dictionaries and college catalogues. I think they do more harm than good because they separate things that should not be separated; because, however useful the words may have been in the past, they have now become handicaps to the further development of knowledge.

Words like botany and zoology imply that plants and animals are quite different things. They are different at one level, to be sure: anyone can tell a horse from an oak tree. But the differences rapidly become blurred when we start looking at the world through a microscope. There are many microorganisms that botanists claim to be plants and zoologists to be animals, with equal plausibility. It was once logical to divide the objects in the world into three great classes—animal, vegetable and mineral—but this distinction now is useful only in parlor games.

The similarities between plants and animals became more important than their differences with the discoveries that both were built up of cells with common basic characteristics, that plants, like animals, had sexual reproduction, and that their needs for nutrition and respiration were similar; and with the development of evolutionary theory, which showed that plants and animals were governed by the same kinds of evolutionary forces. Both are organisms. Unfortunately, organism remains a rather strange and pedantic word that has not really penetrated our basic vocabulary.

Plants and animals are different, of course. When we think of plants we first think of organisms with chlorophyl, organisms able to build up starches from carbon dioxide and water by using the energy of sunlight; they are the basic organisms in the economy of nature, on which all else depends. We also think of plants as fixed organisms; of animals as active, moving ones. And we think of plants in general as absorbing water and food; of animals as ingesting or "eating" it. There are exceptions to all of these. The fungi (mushrooms, molds and the like) are clearly plants, but they lack chlorophyl and depend on other plants as much as animals do. As for movement, the slime molds, generally classed as plants, do a deal of creeping; and many kinds of animals in the sea, like the corals, are as immobile as any tree. I can't think of any plants that gobble their food, though the so-called carnivorous plants trap insects and digest them at leisure. Many animals, chiefly parasites, absorb food in a plantlike way.

These difficulties of definition are trivial. The most serious difficulties can be avoided by dividing all organisms not into two kingdoms, but into three—microbes, plants and animals, because the microbes, in many ways, form a world just as distinctive as that of the visible plants and the visible animals.

But I still would object to dividing the study of living processes into botany, zoology and microbiology because, by any such arrangement, the interrelations within the biological community get lost. Corals cannot be studied without reference to the algae that live with them; flowering plants without the insects that pollinate them; grasslands without the grazing mammals. And at a different level, all protoplasm, all living stuff, shows much the same behavior: the problems of maintenance, growth, differentiation, reproduction, adaptation, evolution, are common to all life and can be studied most conveniently sometimes with one kind of organism, sometimes with another. The differences come out chiefly at the level of classification and cataloguing. This certainly is important enough, but it should not be allowed to swamp all other aspects of the study of life.

The case against physiology and anatomy is somewhat different. Anatomy is concerned with structure, physiology with function—which brings up the very old problem of the relation between structure and function, between how a thing is built and how it works. It is interesting that the first use of the word biology in the English language (it has an older history in German and French) was in a book published by an English surgeon in 1821 deploring, not the separation of zoology and botany, but of anatomy and physiology. And this problem is still with us, only partially solved by labels like "functional anatomy."

Biology shares with medicine the tendency to give every possible kind of specialization a distinctive Greek-root label. This can be understood in medicine because the patient is presumably impressed by the thought that he is in the hands of an otolaryngologist or ophthalmologist or some other variety of learned specialist. It's part of the general medical love for big words and ritual, which I suspect has deep roots in our culture. The quacks and cultists have it as much as physicians in the tradition of scientific medicine. But this does not explain why people who study birds should want to be called ornithologists; insects, entomologists; grass, agrostologists; fungi, mycologists; and so on *ad infinitum* and *ad nauseam.*

I am not trying to argue against specialization. This, in the world of modern knowledge, is necessary. But generalization is also necessary if we are to fit our jigsaw pieces of information together into meaningful patterns. For this, I think the specialist should always be conscious of the relations of his particular subject and his particular point of view to the larger universe of knowledge, which is made more difficult if each way of specializing is called off as a distinct and independent science. There is an element of word magic here: entomology and limnology sound more like things-in-themselves than do insect biology and fresh water biology.

It is sometimes said that the unifying element in all of biology is the cell, that cells are the basic units of biology in the sense that atoms are the basic units of chemistry. I am dubious about this. Cellular organization, to be sure, is fundamental in visible organisms (plants and animals), but the world of the microbes is something else again. There has been a long argument in biology about whether these should be called "single-celled organisms" or "organisms without cells." This is more than a war of words; it involves a whole attitude. If one looks at an amoeba, for instance, as something corresponding to a white blood cell in man, the amoeba becomes "simple," "primitive" and the like. This, in the long run, has not been a very useful way of looking at amoebae, and nowadays I think most biologists refer to them as "acellular" rather than as "unicellular." The amoeba meets the problems of nutrition, respiration, coordination, reproduction, without resorting to cellular differentiation. The amoeba, far from being simple, is quite remarkably complex. But if the amoeba is acellular, we have lost cells as our basic biological unit.

Then we have the problem of viruses. Everyone knows about viruses: they cause polio, influenza, measles and all sorts of other diseases. If the doctor doesn't know what is the matter with you, he is fairly safe in saying, "It's probably a virus." Yet everyone is puzzled about viruses. They certainly have no cellular structure in the ordinary sense. Some of them act like chemical solutions: they can be crystallized, liquefied again, and still show the properties of being alive. They are particulate and it can be shown with the electron microscope that the different kinds of viruses come in different shapes and sizes. Some of them, at least, may be nothing more than giant molecules of a chemical stuff called nucleoprotein. One could plausibly argue that these nucleoproteins are the basic units of life.

But with this we have moved from biology to chemistry. Biology, chemistry and physics form, or some day will form, a single, interdependent system of thought. People who call themselves chemists or physicists will probably eventually go furthest in revealing those tantalizing "secrets of life." But there are still plenty of things to be learned at the biological level, which is our concern in this book. At this level, it seems to me that the most significant and general unit is not the cell, but the individual.

Living stuff is universally distributed in discrete packets, organized in the form of separate, individual organisms. These range from the single molecules of some virus particles, through all the varieties of noncellular organization of the microbes, to the coordinated aggregations of cells of an elephant, a whale or a sequoia tree.

Yet there are difficulties with this concept of the individual. It is easy enough to identify individuals in a room full of people—to tell where one stops and the other starts. One can similarly identify individual marigolds in a flower bed, or individuals among the amoebae crawling across a microscope field. But it is not always so easy: the individual packets of life often maintain organic connections as colonies or clones, and I suspect it is sometimes a matter of rather arbitrary definition whether we call a given aggregate a "colony" or an "individual." We consider a coral clump to be a colony of individual polyps building a common skeleton, but a sponge is an individual organism. The tiny flagellate protozoans that are called *Volvox* form little spheres of dozens of cells that go tumbling along in a nicely coordinated fashion, but we call each sphere a colony. And what about a clump of bamboo or bananas, or a running grass?

There are other difficulties with the idea of the individual when we look at life over a span of time. All life is distributed in discrete packets; but all life is also continuous—the packets are momentary aspects of an ever flowing stream through time. The amoeba or the bacterium splits in two. The protoplasm goes on and we rather arbitrarily decide that the old individual is gone and that we have two new ones. With sex, we get something new: continuity depends on a fusion of parts from two antecedent individuals, and the outward manifestation of the organization becomes discontinuous. The organization, the individual, must die and the continuity is through a germ-cell fragment that, for development, must unite with another fragment. Where is the individual here? I would say that the new individual starts at the moment when the two germ cells fuse—but this again could be called an arbitrary matter of definition.

Yet I think the individual is the nearest thing to an "objective" category that we have in biology. If I were planning a scheme for specialization within the biological sciences, I would take as my point of departure this idea of the individual. My first subdivision would depend on whether interest lay primarily in events inside the individual or outside. I would distinguish, in other words, between "skin-in" and "skin-out" biology.

This has a certain logic because events inside the organism have to be studied by different methods from events outside the organism. Some people tend to be most interested in what goes on inside, in anatomy, digestion, circulation and the like; others in what the whole animal or plant does, how it behaves, how it is distributed around the world, how it lives. This book is about "skin-out" biology, about what I like to call natural history. If you must give it a label derived from the Greek, ecology is as close as anything.

To be sure, events inside the skin are by no means independent of those outside. An animal's behavior toward food, for instance, will depend on whether the animal is hungry or not, and hunger in turn depends on things going on in the digestive and nervous systems. All animal behavior depends to a large extent on sense perception—you can't react to a thing unless you can see it, smell it, taste it, feel it or perceive it in some other way. The study of behavior thus involves the study of perception and sense organs, as well as nerve co-ordination, endocrine gland secretions and all sorts of other things inside the skin. Inside and outside events in a plant are similarly related: you can't study growth, flowering, soil preference, cold tolerance or anything of that sort without taking into consideration what is going on both inside and outside the plant. But this isn't saying that the skin, cortex, membrane or whatever you want to call the boundary of the individual, is meaningless. It is saying that, for many things, you have to look on both sides of the boundary.

"Skin-in" biology, then, is primarily concerned with what makes the individual work, with the functioning of the different organ systems and with the ways in which they are built up of tissues and cells. "Skin-out" biology starts with this individual and is concerned with its relations to other individuals and to the varied aspects of the physical environment. If we start with the individual and go inside, we find ourselves concerned with such units as organs, tissues, cells, molecules. If we work in the other direction, we find that individuals of the same kind form populations, and that these populations can be grouped together into biological communities. When we talk about the economy of nature, we are talking about relationships among populations and individuals within these biological communities—and this is what I want to discuss in this book.

The most complicated, the most highly developed, of natural economics are those of the tropics—of the tropical forests and the tropical seas. These have the largest number of different kinds of organisms living within them, and the many kinds of organisms mean many kinds of relationships. These forests and seas, to the casual visitor, appear to be utterly unrelated places, with nothing in common but the fact of life. Yet when one stops to look, all sorts of similarities, all sorts of analogies, begin to appear.

Since I was first deeply impressed by the unity of the natural world while living in the tropics and working in the rain forest at the headwaters of the Orinoco River in South America, where I was studying jungle yellow fever, I shall start with an account of this experience. I will then shift from the specific to the general and discuss the entire biosphere—the thin film of living stuff that covers the surface of our planet. Perhaps we can better realize both the diversity and the unity of this film of life if we then examine some of its more striking aspects—the life of the open seas, of coral reefs, of fresh water, of tropical forests, and of woods, grasslands and deserts.

With this descriptive background, we shall then turn to the general principles of the organization of biological communities and look at some of the patterns of relationships that appear among individuals, populations and species.

Where does man fit into this system of nature? In the last chapters of the book, I will consider the question of man as a part of nature, and of man as a very special phenomenon, in many ways apart from nature. I cannot make any very clear statement about man's place in nature, though my thinking is undoubtedly colored by my belief that man is a natural, rather than a supernatural, phenomenon. But whatever our beliefs, we are living with nature. And I think we can live more fully, more pleasantly, more productively, if we try to understand the world of nature. And in trying to understand nature, surely we also get new insights into ourselves.

2. Landscapes–
and Seascapes

But there is a growing pleasure in comparing the scenery in different countries, which to a certain degree is distinct from merely admiring its beauty. It depends chiefly on an acquaintance with the individual parts of each view: I am strongly induced to believe that, as in music, the person who understands every note will, if he also possesses a proper taste, more thoroughly enjoy the whole, so he who examines each part of a fine view, may also comprehend the full and combined effect.

—CHARLES DARWIN, in *The Voyage of the Beagle*

For eight years I lived in a small town in the interior of South America—Villavicencio, in eastern Colombia. The town lies at the base of the Andes, where the mountains give way abruptly to the great plains of the Amazon and Orinoco river systems: plains which are covered with grass to the north—the llanos of Colombia and Venezuela—and with forest to the south—the Amazon forest.

It is paradise for a naturalist. Almost every sort of tropical environment can be found within a few miles of the town. North and east are the savannas—the llanos—with long corridors of gallery forest along the streams and rivers. The proportion of grassland slowly increases as the mountains are left behind, until finally there is nothing but grass, the vast empty plains of Arauca. But along the base of the mountains the forest is continuous, and to the south it spreads ever farther out into the plains, making an unbroken canopy for thousands of miles. Most travelers know this Amazon forest only as it stretches interminably beneath their airplane, or as it slips by day after day as they make their way down the major rivers.

Behind the town are the mountains. The first ridge reaches some five thousand feet within a mile or two in a straight line, but it's ten miles as the road winds back and forth across the slopes. There the clouds hung almost continuously, making either rain or fog, and they changed the character of the forest. The trees were not as high as in the plains below, and they were loaded with epiphytes—the many kinds of plants that, in this continual dampness, could grow clinging to the trunks and branches of the trees. The trunks were buried under mosses and ferns and the special epiphytes of the tropics: orchids in endless variety, and bromeliads, the plants of the pineapple family.

The cloud forest was not always damp gloom. When the sun shone, it was especially clean, bright and intense where it penetrated the forest canopy, which was broken in places by cascading mountain streams. The sometimes startling flowers of the orchids and bromeliads seemed particularly bright against the mossy background. And butterflies, here in the cloud forest, were everywhere. Some were plain and dull, but others gleamed in the sun with metallic blue or green, or arrogantly displayed patterns of loud reds and yellows.

Our road to the capital, Bogotá, twisted through the mountains for slightly over a hundred miles—six hours' driving time. The road dipped down again, after the cloud forest, into a narrow valley cut off from the rain-giving clouds by the high mountains on either side. There cacti and thorny shrubs grew in a tiny bit of desert. Then the road climbed up and up, finally to reach the pass at eleven thousand feet, to enter still another world, the paramo.

Some botanists call this kind of country "elfin woodland," which seems to me particularly appropriate. The sun here can be warm and friendly, but it is always cold in the shadows and the landscape is usually wrapped in clammy fog. It is a landscape of miniatures, of gnarled, twisted, lichen-covered shrubs, of fine turf, studded with small, delicate, bright flowers.

We had everything in Villavicencio except the sea. The sea

was far away. The mud from our streets, washed into the Guatiquía River where it passed the town, would have to travel about 2500 miles down the Meta and the Orinoco before it dropped into the delta of the Orinoco or stained the Gulf of Paria off Trinidad. The sea was far away in Villavicencio, but curiously I spent a deal of time there thinking about it, reading about it, wondering about the similarities, the biological analogies, between the forest and the sea. There I first began to realize the essential similarities in plan and function among all the diverse living landscapes and seascapes of our planetary surface—the essential unity of the living world.

I remember very well when my preoccupation with this first started. We were in Villavicencio to study jungle yellow fever, a form of the disease that had been recently discovered in remote parts of South America. Villavicencio was chosen for the study because it was relatively accessible—a road had been built across the Andes in 1933—and elaborate equipment needed for modern disease study could be brought in. A laboratory was built there in 1935, and studies, supported jointly by the Colombian Government and the Rockefeller Foundation, were continued until 1948. I became director of the laboratory in 1940, coming from Egypt, where I had been working on a very different disease, malaria.

Very soon, both in Brazil and Colombia, scientists began to suspect that the disease was transmitted to man by a gaudy day-flying forest mosquito called *Haemagogus capricorni*. It was also found that this mosquito, conspicuous because of its bright, metallic blue color, would appear suddenly in considerable numbers whenever a tree was cut down—and jungle yellow fever could almost be called an occupational disease of woodcutters.

We discovered that this mosquito was ordinarily most abundant high in the trees and that it came down near the ground in numbers only at the forest margin, or where trees were being felled. To study it, then, we needed to get into the treetops.

We tried many ways of doing this. In the end, we made

permanent installations in several forest areas, building ladders up the trees and making comfortable platforms at various heights where we could put instruments and watch what was going on. We set out to analyze the structure of the forest, and we kept up regular observations of the animals and the environment for several years.

Scientists visiting South America often came to see our laboratory and our forest stations. It was while showing the forest to one of these visitors, a botanist, that the idea of the biological similarity between the forest and the sea first occurred to me. We had lunch beside a little stream in a ravine where the forest was dark and quiet, singularly lifeless. Later I persuaded the botanist to climb the ladders on one of the trees, a dizzying experience avoided by many of our visitors. We rested at the 14-meter platform and then climbed on to the 24-meter platform, which was well up in the forest canopy at that particular point.

It was a sunny day, and the contrast between the forest floor and the forest canopy was striking. A nearby tree was in flower and humming with insect life. Mosquitoes, which had been scarce on the forest floor, began to annoy us. We had a good vantage point for bird watching and there were several birds about, including a large hawk perched on a nearby branch, completely indifferent to our presence.

In the course of our mosquito studies we had found that each different species had its characteristic flight habits. Some kinds were found only near the ground, others only high in the trees; some that were most common high in the trees in the morning or afternoon would come down near the ground during the midday hours, showing a sort of daily vertical migration.

While I was explaining this to my friend, it struck me, that this is just the way animals act in the sea. Most life is near the top, because that is where the sunlight strikes and everything below depends on this surface. Life in both the forest and the sea is distributed in horizontal layers.

The analogy, once thought of, was easily developed. The vocabulary for life in the sea could be transferred to the forest. In the treetops we were in what marine students call the pelagic zone—the zone of active photosynthesis, where sunlight provides the energy to keep the whole complicated biological community going. Below, we had been in the benthos, the bottom zone, where organisms live entirely on second-hand materials that drift down from above—on fallen leaves, on fallen fruits, on roots and logs. Only a few special kinds of green plants were able to grow in the rather dim light that reached the forest floor.

My mosquitoes acted in some ways like the microscopic floating life of the sea, the plankton. Each species among the plankton organisms has a characteristic vertical distribution: some living only near the surface, others only at considerable depths, and so forth. The plankton organisms in general show a daily vertical migration, coming to the surface at night and sinking during the day: a migration to which my mosquitoes were only a feeble counterpart. But insects on land are only partially analogous with the plankton of the sea. A major portion of the plankton consists of microscopic plants, busy using the energy of the sun and the dissolved carbon dioxide of the water to build up starch and thus provide the basis for all the rest of the life of the sea. These microscopic plants would correspond not to the insects of the forest, but to the leaves of the trees. The forest insects would correspond only to the animal component of the plankton: to the copepods and tiny shrimp and larval fish which live directly on the plants or on each other at the very beginning of the endless chain of who eats whom in the biological community.

The real basis of the analogy is that both the forest and the sea are three-dimensional. The students of the sea have always been keenly aware of this, but the students of the forest have paid less attention to problems of depth. Of course the scale is utterly different. To compare the "gloomy depths of the forest" with the "gloomy depths of the sea" is so far-fetched as to be ludicrous—though each phrase is apt enough in its own context. The analogy is closest if we compare shallow tropical seas with tropical forests, especially with the great rain forests of the Amazon, the Congo and southeast Asia.

Man's point of view is curiously different in the forest and in the sea. In the forest he is a bottom animal, in the sea a surface animal. To study the forest man must climb; to study the sea, dive. I often thought about the differences as I struggled with the problem of understanding the Villavicencio forest, and in later years as I swam around reefs in the Pacific or in the West Indies. Man is a land animal; it seems probable that much of his evolution took place in the tropical forest, or in regions where the forest was giving way to open grasslands. Yet it seems to me that in some ways modern, scientific man has learned to cope with the sea better than with the forest.

With the invention of the aqualung and similar devices man has gained a freedom in the sea (at least in the top hundred feet or so) that has no counterpart in the forest; but even before this I think diving was easier for him than climbing. And scientists have been more ingenious in developing methods of sending recording apparatus and traps down into the sea than they have in studying conditions up in the forest. In my forest I was always confined to the trunks of the trees, I had no way of getting out into "interarboreal space"—I could be a poor sort of a monkey, but I had no way of being a bird. In the sea I am one with the fish; I can float or dive or pause suspended in their midst; I must live in a different way, my dreams of floating through the forest, with the birds. Perhaps that is why I so much prefer fish watching to bird watching.

But let's look a little more at the biological similarities between the forest and the sea. Both have a vertical gradation in light, with associated gradations in other aspects of the environment, such as temperature, and air or water movement. This is due in one case to the arrangement of vegetation and in the other case to the properties of water itself. And both environments are relatively stable, in one case because of the insulating effect of the mass of vegetation and in the other because of the density and heat conservation of water. The physical properties of water underlie the insulation and stratification in both environments—though in one the water is free and in the other tied up in the protoplasm and sap of the forest trees.

The climate of the floor zone of the tropical rain forest is very constant and uniform. The temperature difference between midday and midnight in our Villavicencio forest was usually only one or two degrees, and the difference between the coldest month and the hottest month was only 4° C. The humidity was similarly constant in the lower levels of the forest, where the air was usually saturated with moisture, and the relative humidity rarely dropped below 80 per cent even at midday in dry weather. Air movement within the forest was slight. This was noticeable in taking photographs: five- or ten-second exposures were often necessary because of the dim light, yet there was no trouble because of leaf movements. Sometimes in the dim and quiet depths of the forest you could hear the wind rustling through the treetops, though the air below seemed not to stir. The plants that grew on or near the ground in the forests often had large, delicate leaves that would have been torn and tattered by any wind—the sort of plants that we admire in hothouses. They were comparable with the delicate and fantastic growths in the quiet depths of a tropical lagoon.

The daily fluctuations in climate, then, are least near the ground in the forest, greatest in the open air above the canopy. At midday the temperature is lowest near the ground, the air becoming warmer as one climbs into the canopy; the humidity is highest near the ground, and the light, least. At night the temperature gradient is reversed, the air being warmest near the ground. Relative humidity at night is near saturation at all levels in the tropical forest, so that differences in this respect are measurable only during the day.

The gradients in light, temperature and air movement correspond with similar gradients in the sea, though the scale is of course very different. Moisture gradients, so important in governing the distribution of plants and animals on land, have no counterpart in the environment of the sea.

It is interesting to compare the ways in which organisms keep their places in the forests and in the sea. The sea has many kinds of animals and plants that grow in fixed places, and these sometimes make up a considerable mass, as in beds of giant kelp or in coral reefs. In general, however, life in the sea depends to a very large extent on floating and swimming. Flight is only partially comparable with swimming because it takes much more energy for an animal to support itself in air than in water. Seeds, spores and a few animals like spiders that cling to a long thread of silk truly float in the air, but this is nothing compared with the life that floats in the sea. Some of the butterflies of the tropical forest have achieved a leisurely, soaring flight whereby they seem to maintain themselves in the air with almost no effort, but they must alight from time to time, which an animal in the sea need never do. In general, the orientation of life in the forest is dependent not on flying or floating, but on adaptations for climbing and perching in the trees. Trees, forming a network of trunks, branches, twigs and leaves, play the role in the forest that corresponds to the supporting role of the water in the sea.

Life reaches its greatest diversity in tropical seas and tropical forests. Warmth, light, moisture, the three essentials for life, are here always present and dependable. It is generally thought that life got its start on this planet, a couple of billion years ago, in warm, shallow seas; and these seas continue to provide the most favorable environment for life that we can imagine. Animals and plants have really never learned how to leave the sea. In going on land, they have learned how to take a bit of the sea with them. In a sense all the land organisms are packages of sea water, variously wrapped and supported. But all these clever packages "leak" a little, and they must have some way of getting water back. This is easiest in the wet tropics, where the water supply is steadier and more dependable than in any of the other land environments.

We could see the effect of moisture easily enough without going very far from Villavicencio. The pattern of forest and grassland in the plains reflected the pattern of the availability of water. The band of rain forest along the base of the mountains lay in the zone where the clouds, striking these mountains, first dropped their moisture. The rain was not uniform through the year: there was much more in April and November than in August or February. We were just four degrees off the equator, so the sun passed back and forth overhead twice a year with a slight, but measurable, consequence in seasonal changes. The period from the end of December to the middle of March was the driest, but there was always some rain.

As one left the mountains and moved to the east or north, the annual amount of rain dropped considerably. Even more important, perhaps, the dry periods became drier, so that fifty miles out from the mountains, January or February might pass with no rain at all. In the uniform warmth of the tropics, evaporation is rapid and organisms need a much higher rainfall to maintain themselves than in cooler climates. Consequently, as one moves away from the mountains, the forest becomes restricted more and more to the banks of the rivers and streams—to gallery forest, supported by ground water and flood water—with the intervening country covered by grass.

Unfortunately, simple explanations are rarely complete explanations. In general, one can say that the forested regions of the world are regions of high rainfall, the grasslands, of low rainfall. But in studying grasslands in almost any part of the world we have to take into account another factor, fire. Every January or February the grasslands of northern South America are fired, now mostly by the ranchers, who thus prepare for a fresh growth for their cattle. But it seems that long before the Europeans penetrated this part of the world, the

Indians had the custom of firing the savannas. And ecologists argue about whether, before the Indians, the fires were not often started by lightning.

Whatever the history, it is clear enough that the present landscape is governed by the combination of rainfall and fire. The long dry season makes the firing possible, but the repeated firings mean that only fire-resistant plants can remain established. The rain forest itself never becomes dry enough to burn except where man cuts down the trees and fires the dead foliage—the slash-and-burn agriculture of farmers everywhere in the forested parts of the tropics. Where the trees have not been cut, the fires sweep up to the edge of the forest and stop. If fire could be eliminated, I suspect that the grasslands would become covered with a rather different sort of vegetation, perhaps a tangled shrub. But no one has tried the experiment and I can only be sure that whatever the vegetation, it would not be the forest that we know in the zone of high rainfall.

If, from Villavicencio, we move up the mountains instead of out into the plains, we find the effects of both moisture and temperature. The cloud forest, with its constant fog, is certainly different from the rain forest. It does not have the cathedral-like magnificence of the rain forest, but I would hesitate to say that the conditions for life there are more unfavorable. The temperatures are lower, but at a level, say, of five thousand feet near the equator they are not low enough to have a general, unfavorable effect on animals and plants. The cloud forest has a fantastic variety of living things, of orchids and insects and ferns and trees. I don't know that anyone has made a count, or even an estimate, but I suspect that the cloud forest at five thousand feet near Villavicencio might well have as many different kinds of animals and plants as the rain forest below. But this cloud forest is a very special, restricted kind of environment, found only in rather narrow bands or occasional valleys in the mountain ranges of the tropics. There are no vast stretches of cloud forest comparable with the interminable rain forest of the Amazon or the Congo.

The effect of temperature soon becomes apparent as one continues to climb in tropical mountains. All life, as I've said, consists of packages of water, and water turns solid at 0° C. or 32° F., which makes for special problems when these temperatures are reached. Many organisms, obviously, have found ways of dealing with this problem, but it remains a special problem, and the number of kinds of animals and plants that can survive drops rapidly as one climbs mountains in the tropics or goes toward the poles outside of the tropics. Finally, in the icecaps of tropical mountains or of the polar regions, one has passed beyond the limit where any life can maintain itself.

On land, then, we can look at the distribution of life on the surface of our globe in terms of temperature and moisture, and find a nice series of conditions from the most favorable environment of the rain forest to the completely barren conditions of the mid-Sahara or mid-Antarctica.

When we look at the sea, moisture no longer matters, but light becomes important, and temperature takes on somewhat different meanings. Different wave lengths of light penetrate sea water to different depths, and the degree of penetration also varies greatly with the clarity of the water and the angle of the sun. Any broad statement is thus bound to have all sorts of exceptions and modifications. But light always gets cut down as depth increases in the sea, and always, within a few hundred feet of the surface, one is in a region of complete blackness. Thus life at the surface of the sea is necessarily completely different from life in the depths.

This has no counterpart on land. I can play with my anology between the forest and the sea, and I still think it is helpful and illuminating, illustrating some general principles of living things, but the great depths of the sea form a special and unique environment, mostly inhabited by very special organisms.

In the depths it is always dark and always cold, regardless

of geography. Only the surface layers of the sea show neat geographical patterns like those of the land. Here, near the surface, we find a special exuberance in the tropics, and the coral reefs form an environment even more gaudy and wonderful, from the point of view of intruding man, than the rain forest. The seas, because of the properties of salt water, form everywhere an environment more stable than any on land, freezing only locally and in high latitudes. There is a dropping off of kinds and types as one moves from the tropics toward the poles, but the difference is less sharp than on land. And it can be argued that in terms of productivity of individuals, of teeming codfish or crustaceans, the northern seas are more productive than the tropics. On land too, of course, the difference in the number of kinds of things is more striking than the difference in the number of individuals: a northern forest may have as many trees per square acre as a tropical forest, but the one will consist of one or two or three kinds of trees, the other of hundreds of kinds.

But before we look at this in any detail, it may be well to consider some of the general aspects of the living world. Rain forests, cloud forests, coral reefs, ocean depths, all these are examples of different kinds of biological communities. These different communities, however, are not separate and independent entities; they are interrelated parts of the total system of the world of life, of the biosphere.

3. The Living World

. . . the discoveries of Darwin, himself a magnificent field naturalist, had the remarkable effect of sending the whole zoological world flocking indoors, where they remained hard at work for fifty years or more, and whence they are now beginning to put forth cautious heads again into the open air.

—CHARLES ELTON, in *Animal Ecology*

The idea of the continuity of life is, in one way, a very old one. Arthur Lovejoy, in a book well known to scholars, *The Great Chain of Being,* has traced the history of the idea from the Greeks through medieval philosophy to modern times: the idea of an unbroken chain of living forms, from the lowly crawling things of mud and slime through fish and lizards and birds and mammals to man, midway on the scale of things toward the ever increasing perfection of the heavenly hosts culminating in God himself. Every possible kind of thing, by this philosophy, exists; every niche is filled, each organism has its appointed role.

Yet in another way, the idea of the continuity of life is quite new, the product not of philosophical speculation, but of scientific analysis. I don't want to get sidetracked here into the difference between the philosophical and scientific points of view —a difference perhaps more apparent than real and certainly in itself a product of the modern mind. The difference exists, however tenuous or artificial its basis; and we can date the scientific awareness of the continuity of life from the beginning of the Nineteenth Century.

This continuity has aspects in space and in time. The idea of continuity in space is covered by the word biosphere; in time by the word evolution. The word biosphere was invented by the French biologist Jean Baptiste de Monet, best known as the Chevalier de Lamarck, in 1809, to denote the whole zone at the surface of the earth occupied by living things. The word evolution has a less precise history, though it is notable that Lamarck was also among the first biologists to propose a general theory of evolution. We tend to associate the word with another man, Darwin, and with the date of his *Origin of Species,* 1859, because it was Darwin's theory of evolution that first gained general acceptance among the scientific community.

I would like to postpone discussion of evolution, of continuity in time, and look first at the biosphere, at the continuity of life in space. The living world of the biosphere, the thin but continuous film on the surface of our planet, is held there because only at the surface are the chemical and physical conditions appropriate for life's development. The film of living things is thickest in the seas, because it is now clear that some forms of life have been able to penetrate to even the greatest depths of the oceans. On land, however, the film is very thin indeed, confined to the few feet of surface soil which air and water can penetrate and, essentially, to the height above the soil reached by the tallest trees. Birds, of course, can soar high into the atmosphere, and with traps on airplanes it has been found that insects and the like may be carried upward thousands of feet on air currents. But life cannot sustain itself unattached in the air: the birds, the insects and the spores have escaped the surface only temporarily and to survive must come back to earth from time to time. Below ground, some very special forms of life have been able to penetrate deep caves and to live in underground streams, but these caves and streams represent, in a way, intrusion of certain surface conditions into the depths of the earth. Even if we allow for the deepest oceans and deepest caves, and for the greatest heights of flying or floating into the atmosphere, we are dealing with a layer of a few thousand feet on a planet measured in thousands of miles. The thickness of the biosphere is essentially the same as the thickness of the irregularities of the earth's surface.

The biosphere, to a casual glance, is varied enough, broken into oceans and continents, into forests and grasslands, rivers, lakes, bays, shallow seas and abyssal depths. The essential unity of the biosphere, however, emerges when we start to classify these variations. Every classification becomes in some sense arbitrary since it is an attempt by the human mind to impose categories on a continuous system. The differences are real enough when we compare different parts of the system: a forest with a pond or with a coral reef. It is only when we try to establish precise boundaries, neat definitions, that the differences become blurred and start to seem unreal. This is because we are not dealing with discrete, independent entities. We are not counting apples and pears and oranges; we are trying to measure the garden with a yardstick with no agreement about the size of our inches.

There is little argument about the three basic subdivisions of the biosphere: the seas, the land and fresh water. The conditions of life are quite different in each of these environments; by and large, each has its characteristic inhabitants, each is in turn divisible into an endless series of living communities differing in details yet with basic similarities. But when we approach the boundaries, the differences between even these basic subdivisions of the living world become blurred. The sea margin everywhere illustrates the blending of the marine and terrestrial environments: mangrove swamps, tidal flats, sandy beaches, all such habitats include a mixture of inhabitants from the land and the sea. The difficulty in separating the two under some circumstances, however, became most apparent to me during a summer spent in studying life on a tiny coral atoll in the western Pacific, a study which Donald Abbott and I have reported in *Coral Island.*

Our atoll consisted of a circular reef in the open Pacific enclosing a lagoon about a mile in diameter. The reef, for the most part, was barely awash at low tide, but in places the coral had piled up to form islets; the total land surface of these islets was a half a square mile. There were 260 people on this atoll, living on the breadfruit, coconuts and taro that they grew on land, and on the fish they got from their lagoon and from the surrounding sea.

Here on the atoll of Ifaluk, the distinction between land and sea seemed to lose its biological meaning. We could find no logical way of subdividing the environment into a series of discrete biological communities and we came to the conclusion that the meaningful community included the whole atoll situation: land, reef, lagoon and immediately surrounding sea. Everything was all mixed up. The people depended equally on the land and the sea for their food. The hermit crabs that crawled everywhere over the atoll went to the sea to lay their eggs, as did the coconut crabs and the land crabs; the sea turtles came out to bury their eggs on land. The influence of the sea was everywhere; it determined what plants and animals were living on the land, because all of these, to get there, had to have some method of crossing the sea, unless they had been purposefully or accidentally brought by man.

We could, of course, analyze our atoll in terms of biological subdivisions such as coconut grove, taro swamp, boulder ridge, beach, intertidal zone, lagoon patch reef, outer reef margin, and the like. Some items in such a list would be wholly marine, some wholly terrestrial, some mixed, but none were biological *communities* in the sense of being relatively independent, relatively understandable as entities in themselves. Food relations, reproductive relations, dispersal relations, constantly crossed the boundaries of these subdivisions. Our only intelligible field of study was the atoll environment as a whole.

Land and fresh water similarly blur in many situations. A

host of different kinds of insects spend part of their lives in fresh water, part on land, as do many kinds of birds and mammals. Words like pond, marsh, swamp, bog, illustrate the problems of sharply separating land and water situations for either plants or animals. Similarly in the estuaries of rivers and in coastal marshes the life of the sea blends with the life of fresh water.

The biosphere, then, is essentially continuous in space, a single interwoven web of life covering the surface of our planet. But it is far from being a uniform, monotonous web: it is woven into a motley series of patterns and designs. The biologist has, on the one hand, the problem of finding the uniformities, the underlying motifs common to all of the patterns; and on the other hand, of finding and describing the divergencies, the meaning of the variations, the characteristics of the separate designs that have gone into the making of the web.

Let's first look at divergencies, at some of the different designs that appear in this web of life.

Of the three major subdivisions, the sea is by far the largest, the oldest and the most continuous. Something like three-fourths of the surface of the planet is covered by sea; and while the land masses break this sea into a series of oceans —Atlantic, Pacific, Indian—in middle latitudes, farther south the waters are continuous. Any animal of the open sea that can withstand the cold of the south polar oceans is thus free to wander everywhere in this area. Geography has little meaning except for organisms with narrow temperature tolerances, or for organisms chained to the margins of the lands.

We presume that life started in the sea and that for a very long time it existed only in the sea. This is a presumption. We have no absolute proof, and perhaps we shall never have, because no one was there to see and the few clues may always be difficult to interpret. For a long time the origin of life was not considered a problem; life seemed to arise spontaneously, constantly, all about us—in the weeds that shot up in the spring, the maggots that appeared in dead flesh, the mice that were bred by filth and waste. The gap between the living and the dead in the world around us has always been far from obvious.

The idea of spontaneous generation died slowly. The Italian, Redi, showed long ago that maggots appeared in meat only when flies could get at it to lay their eggs; but then, with the invention and perfection of the microscope, the world of microbes was discovered, and these again seemed to rise spontaneously whenever conditions were appropriate. Pasteur, with his dramatic experiments, gets the credit for showing that these microbes, too, could come only from other microbes. Life could come only from life; there was no present bridge across the gulf from the inanimate to the animate.

The discovery of the viruses—of life apparently in a fluid form, or in the form of submicroscopic, molecule-like particles—opened questions again. It can be shown easily enough that viruses come only from viruses, that no organism shows a virus disease unless it has somehow been infected with the virus. But all the viruses that we know are parasitic, and we recognize them only through the damage they cause to other organisms; there may possibly be free-living, virus-like living substances that we have no way of recognizing. If you find something, it is easy enough to show that it exists; but it is very difficult to prove that something that might possibly exist does not, in fact, exist.

Thus, while there is always some bare possibility that life may still be forming by some unknown process under some unknown circumstances and in some unknown form in our world today, this possibility seems, to say the least, extremely remote. All our evidence shows that life as we know it is a unity, self-perpetuating: that life comes only from life. Yet somehow, somewhere, sometime, life must have got started. One idea is that life might have arrived on earth from outer space, riding a meteorite, and many meteorites have been overhauled for traces of some possible germ. But even if this were the case, it would simply push the problem out from the earth to some other situation: the problem of beginnings would still be there.

The vast majority of biologists think that life started on this earth and that it started some two billion years ago under conditions vastly different from those prevailing today. The problem, then, becomes one of trying to imagine the conditions that might have prevailed at the very beginning of life, and of trying to reproduce those conditions in the laboratory to see what might have happened.

We are gradually coming to realize how different physical conditions would be at the surface of our planet without life —which means also how different they were at the time that life got started. Life, to be sure, is a consequence of environmental conditions at the earth's surface, but this environment, at the same time, is to a considerable degree affected by the life processes. This is one aspect of a problem that appears over and over again in modern biology: the problem of separating "organism" and "environment." The distinction seems simple and clear enough at first glance, but it breaks down in many different ways—until sometimes I wonder whether it isn't a mistake even to try to make the distinction, whether the whole idea isn't basically misleading.

It is now thought, for instance, that the atmosphere of the earth, before life appeared, probably contained no oxygen and no carbon dioxide: that these gases, which we think of as essential for the existence of life, are the products of life. It has been calculated that the oxygen now present in the atmosphere is renewed through photosynthesis every 2000 years, the carbon dioxide every 300 years. The earth's original atmosphere was probably composed of gases we think of as poisonous, like methane and ammonia. With a different atmosphere, the radiation reaching the surface of the earth would also have been different, with important consequences.

In 1953 a student chemist, S. L. Miller, set up an experiment in which water vapor, methane, ammonia and hydrogen —gases that presumably formed the earth's early atmosphere

—were circulated continuously for a week over an electric spark. Under these circumstances, traces of amino acids—the complex organic building blocks of protoplasm—were found to have formed in the water. One can imagine that the early seas became a sort of thin, organic soup, in which almost anything might happen, including the beginning of some sort of self-duplicating process: the beginning of growth and reproduction, the very essence of life.

It is a long way from an organic soup to the simplest kind of organism that we can imagine, and it may be a long time before we can reconstruct possible steps in this development with any clarity or confidence. But of one thing we are fairly sure, that this development took place in the sea, and that living was confined to the sea for a very long time. Special problems had to be solved before life could get onto land or into fresh water. Life has never escaped from the sea: the organisms of the land and of fresh water carry a bit of the sea environment with them, and maintain the salt and water balance of their protoplasm in the alien environment of air and unsalted water.

Of the three basic subdivisions of the biosphere, then, the sea is the oldest. In many ways it can be considered the most "natural" environment for life, the easiest environment. It contains the greatest variety of life, the greatest number of phyla and classes of living things—not the greatest numbers of species, because a few groups of organisms like the insects and the seed plants that have evolved on land have proliferated enormously into arrays of species adapted to the variegated conditions of the terrestrial environment.

Actually, when one stops to look at it, life on land is built up from a remarkably small number of fundamentally different types of organisms. There are the ferns and the seed plants, the insects, reptiles, birds and mammals. There are, to be sure, plenty of fungi, mosses and lichens; there are many kinds of snails; there are a variety of worms; and there are, everywhere, the unseen microorganisms. But by any accounting this is a poor variety of basic types compared with the life of the sea. The biologist divides all life into some thirty basically different phyla, each subdivided into several distinct classes—the number varies somewhat according to the opinion of the biologist making the classification. The vast majority of these are not found on land at all. Among plants, the endlessly varied members of the several phyla of algae are either missing or represented by inconspicuous species. Among animals, the great phylum Coelenterata – the jellyfish, corals, sea anemones, and the like—is entirely missing, as is the phylum Echinodermata—the starfish, sea urchins, sea cucumbers and sea lilies. On the other hand, no phylum found on land is absent from the sea.

In some respects the life of fresh water is intermediate between that of the land and sea. Some of the marine phyla, like the Coelenterates, have a few representatives in fresh water, even though none have learned to live on land. Others, like the algae among plants and the molluscs among animals, have many more species in fresh water than on land, though still far less than in the sea. More types of organisms, in other words, have been able to solve the problem of keeping their salt balance in the dilute medium of fresh water than have been able to solve the problems of retaining both salts and water on land.

The amphibians—frogs and salamanders—of course lead an existence that is half on land and half in fresh water. And the typically land organisms, like seed plants, insects, reptiles and mammals, have often developed adaptations that enable them to live in fresh water. In general the life of fresh water is closely bound up with the life of land, perhaps in part because the fresh water is so widely dispersed over the land as ponds, lakes, streams and rivers, completely lacking the great continuity of the sea in either space or time. The history of fresh water is tied to the history of the continents more than to that of the sea.

Thus when we look at the biosphere as a whole, we find

an essential unity. When we go on to analyze this, we find we can recognize essential differences among the environments of the sea, the land and the fresh waters, despite the many ways in which these environments, and the organisms living in them, blur. The blurring, the difficulty in establishing sharp distinctions, increases as we go on with our analysis, as we try to describe the characteristics of deserts, grasslands and forests; or as we try to classify the different kinds of forests that we see in the landscapes of any continent. But before going on to take a more detailed look at these landscapes and seascapes, I would like to make a few more remarks about the general characteristics of this living world.

We look at the world, necessarily, as men. This inescapable fact has several sorts of consequences, some more easily realized than others. The most important consequence is that we must always deal with the world in terms of the kinds of symbols with which our minds work—verbal, mathematical, pictorial—and the working of our minds doesn't necessarily coincide with the working of the part of nature that we are observing. But this is a problem for semantics, psychology, philosophy, which can be sidestepped here. In our minds, in our cultures and traditions, in our development of knowledge through the manipulation and storage of symbols, we seem curiously detached from the rest of nature. Yet at the same time we are clearly a part of the natural system. We are vertebrates, mammals, primates; we can be classified as a species and given the name *Homo sapiens*, with or without irony. We are land animals, social animals, in many ways predatory animals, though our dietary habits are not specialized. Life in the trees has left its traces in our vision and bones and muscles, though we are not very good at climbing now. These things all color our view of the world, and it is very difficult to be sure that we have made the appropriate color corrections.

But there is another thing. In the last few thousand years, man has acquired a tremendous power through the explosive development of his cultures. In the perspective of geological time, it is a split-second explosion, a matter not of thousands of years but of milliseconds. With this explosion, man has become a geological force, like glaciation or vulcanism. He not only is capable of altering the landscape and the balance of the biosphere; he has altered them.

If you fly from New York to Chicago, or from Paris to Rome, the surface of the earth below you is obviously man-controlled—fields and highways and towns. The forests of the Appalachians or the Alps may seem, at a glance, as though they were free from human control, but a closer look shows that even there the growth is a purposeful or accidental consequence of human activity. To be sure, if we fly from Port-of-Spain to Manáos across the highlands of Venezuela and the lowlands of the Amazon, we see few traces of human activity; but such regions, in the habitable world, are rare.

It is hard to realize the pervasiveness of this human influence. It is obvious enough around the Mediterranean or in the eastern United States or in China; but many parts of the tropics look wild enough or natural enough. At least they are unkempt. I think I first really felt the pervasiveness of this human influence in the tropics when I happened to wander away from it. I was making a canoe trip down one of the tributaries of the Orinoco into a territory where the only human beings were scattered Indians, essentially playing the part of one more species of animal in the local fauna.

What I had thought of as the "common" plants, the "common" butterflies of tropical America, disappeared. Or at least, instead of being common, they became extremely rare, confined to scattered sand bars on the river, or to rare openings in the forest made where some great tree had fallen. I had to revise completely my ideas about the relative abundance—in biological terms, the relative success—of different kinds of animals and plants. The trees, shrubs, herbs, birds, mammals, butterflies that we see even in wild-looking places are the species that have resisted this human influence, or thrived with it. Getting along in the biosphere has come to mean getting along with man.

This human influence is a tremendously important and interesting thing to study in itself. I am emphasizing it here in part because I think biologists tend to underestimate it, or when they think about it, tend to deplore it rather than study it as a process of biological evolution and adaptation going on at considerable speed under their noses. But chiefly I mention it here because I think it is something we must always keep in mind in looking at the biosphere.

The depths of the sea, so far, have hardly been touched by this new phenomenon of man; he has scarcely started exploring them. But this may change rapidly as he starts using the depths as a dumping place for radioactive wastes. In general, man's influence on the life of the sea has been comparatively trivial, though it may seem important enough if we look at particular animals like whales, or particular places like the North Sea, the Mediterranean coast or any harbor area.

But man is a land animal, and it is on land that we see the effects of his dominance most clearly. I want, at the end of this book, to review the general question of the relations between man and the rest of nature; here I would like only to raise the question, so that it can be kept in mind as we look at particular aspects of the biological world.

Man's direct effects are obvious enough and spectacular enough: the canyons of Manhattan, the patchwork fields of central Europe or Ohio, the dikes of the Netherlands, the drainage system of the Pontine Marshes or the rice terraces of the Philippines. But beyond these direct results of clearing, cultivating and constructing, there are endless indirect effects arising from human activity. Often man has changed the chemical balance of the soil; he is changing the chemical balance of the atmosphere, recently through nuclear explosions, over a longer term through the gases spewed out by his countless chimneys. The fauna and flora everywhere have been altered by the plants and animals that he has moved about, sometimes purposefully, sometimes accidentally. Many kinds of organisms have become extinct as a direct or indirect result of human activity; others, once rare, have come to flourish. He has made fires and floods—perhaps presently he will be making rain.

One tends to think of the major effects of man on the rest of nature as rather recent in human history; certainly many of the most striking developments have come since the global movements of Europeans started in the Fifteenth Century, and with the industrial and population growth of succeeding centuries. Much older influences are obvious in the regions of the ancient civilizations, and painstaking archaeological work in such places as England and Denmark has shown that man started influencing the landscape long before he could be called civilized.

It is hard to say when man started to become a rather special sort of biological force. I suspect that it goes back to his first acquisition of tools and fire. Many curious things happened during the several hundred thousand years of the Pleistocene, the last of the geological periods. Among them was the rise of man, and concurrently the extinction of many other mammal types, including things like the woolly mammoths and the giant sloths. Animal types have reached extinction all through geological history, without any aid from man. But the suspicion remains that man may have had something to do with the last wave of disappearances, particularly since there is direct evidence that many of the mammals, including the woolly mammoth, survived long enough to be contemporary for a while with modern man, with *Homo sapiens*.

But let's try to look at the biosphere without thinking too much about the special case of man—try to see something of the many different kinds of patterns that go to make up the grand design. We'll start with the sea—the open sea—and then look more closely at a special aspect of the sea, the coral reef.

4. The Open Sea

No philosopher's or poet's fancy, no myth of a primitive people has ever exaggerated the importance, the usefulness, and above all the marvelous beneficence of the ocean for the community of living things.

—L. J. HENDERSON,
in *The Fitness of the Environment*

The sea covers most of the surface of the earth—about 70 per cent of it. Furthermore, while the greatest heights of land (Mt. Everest, a little over 29,000 feet) are roughly comparable with the greatest depths of the sea (the Marianas Trench, 35,600 feet), the average depth of the sea is very much greater than the average elevation of the land. If our earth had a smooth surface, the waters of the sea would cover it to a uniform depth of about a mile and a half. As it is, the average depth of the oceans is calculated to be about 12,500 feet, while the average elevation of the continents is about 2500 feet.

Dry land, then, is a sort of accident on the surface of our earth, a consequence of the uneasy balance between the forces resulting in the irregular uplift and constant leveling tendencies of the process of erosion. Yet the basic pattern of the seas and continents, most geologists believe, has remained about the same during the two or three billion years in which life has been developing. There have been tremendous changes in the relations between sea and land, clearly enough marked by beds of marine fossils high in our present mountains. But these changes, we think, represent the extension of shallow seas over present continents rather than any great rearrangement of the continental masses. It seems unlikely that the present continents ever represented ocean depths, or that the present depths of the Atlantic and Pacific were ever parts of continents now lost.

We now realize, however, that sea level itself, through geological time, is not constant: that the volume of water in the seas is not a constant quantity, because varying amounts of the earth's water may be locked up as ice in different climatic periods. We are still living in an ice age, with great continental glaciers on Antarctica and Greenland as well as smaller glaciers in other places. If these glaciers melted—and it seems likely that they are now slowly melting—the added water in the oceans would raise sea level something like sixty feet. This is a trivial addition to the volume of ocean water—but it would make a considerable difference in the details of present coast lines. There would be very little left of Florida; and one wonders at what point Manhattan Island would be abandoned to the rising seas. In the past, sea level has probably varied something like 300 feet between periods of maximum and minimum glaciation.

The varying details of the pattern of land and sea, resulting from uplift, erosion and changes in sea level have had important consequences in the history of land organisms. North and South America have sometimes been connected by land at Panama, have sometimes been separated by the sea; Alaska and Siberia have sometimes been continuous dry land, have sometimes been separated by the Bering Strait. The history of land organisms everywhere can be understood only in terms of these connections and separations and in terms of the climatic changes in the Far North and Far South. But from the point of view of the history of the life of the sea, such changes have been far less important. Mountains and glaciers have come and gone, but the sea has stayed much the same.

Sea water, it seems likely, has had much the same composition for these last two billion years. This is not easy to understand, but sea water is remarkable any way you look at it, and not easily understood. Geologists first assumed that the seas must have started as fresh water and gradually acquired their salts from the slow erosion of the land. With a few measurements and some bold estimating you can figure out that something like three billion metric tons of material from the land are being washed into the sea every year; you can figure out how much sodium or calcium or magnesium is contained in this, estimate how much there is in the sea at present, and thus estimate how long it took to accumulate.

But it turns out that the material being washed into the sea every year has no appreciable effect on the composition of sea water. The salts of the sea, by this system, could be accumulated in only a few million years, while all other evidence shows that the seas have been around for a much longer time. And where salts have accumulated in water from land erosion into lakes without outlets, like the Great Salt Lake and the Dead Sea, the balance of salts in the water is quite different from that in the ocean. The salts of sea water, in other words, are not a mere static accumulation of things washed in from the land. Sea water, rather, represents a balance or equilibrium: materials are constantly added, but materials are also constantly removed. Sea water is the environment of marine organisms, but it is also the consequence of the activities of these organisms. The seas, like the atmosphere, would undoubtedly have a very different composition in the absence of life.

Sea water on the average contains about 3.5 per cent of dissolved salts—a salt content of 35 parts per thousand. This is, in general, the salinity of the open oceans at a depth of about a thousand feet. Surface waters vary considerably around this average, and there are also variations at different depths in the great ocean basins. At the surface, salinity is highest in the tropics, where evaporation is greatest, and lowest towards the poles, due to dilution from melting ice. The Red

Sea, with no inflow of fresh water and with high evaporation, has salinities over 45 parts per thousand; in the Mediterranean, salinities vary from nearly 40 parts per thousand off the Syrian coast to 37 parts per thousand near the Strait of Gibraltar. The Baltic Sea, on the other hand, with a large inflow of fresh water, has low salinities—below 10 parts per thousand in many places.

Common salt, sodium chloride, predominates—accounting for over three-fourths of the dissolved salts. Of the metallic ions, magnesium, calcium and potassium rank next in abundance after sodium, and a great variety of chemical elements have been found to be present in traces. The rare elements are often concentrated by different marine organisms, undoubtedly playing important parts in their biology: some microscopic protozoans, the radiolarians, build their skeletons with strontium; ascidians (curious and remote cousins of the vertebrates) have vanadium in their blood. These elements are hardly detectable in sea water with the most refined chemical techniques, yet the organisms extract them in appreciable quantities.

It is hard to avoid the impression, sometimes, that sea water is a mysterious, magical substance. You can't make it up in the laboratory by putting all the proper chemicals together; something important is always missing, probably some of the rare trace elements, probably also some of the things contributed by the microscopic and submicroscopic forms of life. Every child brought up by the seashore must have been disappointed by his efforts to take some of the sea home in a bucket: he can do this with a bit of a pond or a lake, but with the sea it is much more difficult. The fish and crabs and seaweed live hardly any time, and the water itself seems to decay. The sea is a unit that does not easily allow small bits of it to be carried away by boys with buckets.

The dynamics of the sea have been studied now for many years by scientists: oceanography and marine biology have become flourishing and fascinating sciences. Yet there are still many mysteries about the sea, many factual details of knowledge to be accumulated, many startling discoveries to be made when time, knowledge and technique are ripe. But we have learned a great deal about the sea and our knowledge increases every day.

The economy of life in the sea is based on the plankton. *Plankton* is a Greek word, often translated simply as "wandering." But Greek scholars say that its meaning is more subtle than this, that it carries the passive sense of "that which is made to wander or drift." Wandering and drifting everywhere through the seas are the hosts of the microscopic plankton on which all the rest of the life of the sea depends.

I don't know that any single person can be given credit for "discovering" the plankton: it is an idea that emerged more or less gradually as early naturalists started towing fine nets through the water and discovered the bewilderingly diverse kinds of things that made up the gelatinous mess they had trapped in their nets. Plankton study still involves nets, beautifully fine nets and all sorts of associated apparatus built for the purpose of straining out and concentrating this invisible drifting world of life.

The plankton organisms are usually divided into "plants" and "animals." The animals, mostly microscopic or very small, live directly off the plants and in turn become the principal food of the larger and more conspicuous animals of the sea. The plant component is meant to include all of the organisms that are able to build up starches through photosynthesis, with energy from the sunlight. These microscopic plants play the role corresponding to that of green vegetation on land, and while they are invisible, they are very numerous. Most estimates of "productivity" find that the build-up of organic materials under an acre of sea surface is not strikingly different from that of the vegetation over an average acre of land.

It is not easy to make reliable estimates of the numbers of plankton organisms present in the sea at different places or at different times. It is simple enough to calculate the numbers of organisms in a particular net haul by counting a few samples from the catch and multiplying to get the total. It is less easy to calculate the volume of water from which this catch has been strained, but it is still perfectly possible. The usual method is to mount a small propeller in the mouth of the net which can be calibrated so that its revolutions give a direct record of the volume of water that has passed through the net. But how does one tell how many organisms have escaped through the pores of even the finest nets?

Alister Hardy, in his splendid book *The Open Sea; the World of Plankton,* has reviewed the various attempts to estimate the numbers of oceanic plankton. In one bay in the Isle of Man, regular records of plankton hauls have been kept for many years. By averaging the April hauls for fourteen years, it was found that each cubic foot of water contained about 20,000 microscopic plants, and about 120 plankton animals. This is an impressive enough figure, but it does not include the hosts of very small plant cells that would not be caught by the net. The most ingenious efforts to determine the number of these have been made at the marine laboratory at Plymouth, England, and it turns out that in the ocean water used in these experiments there must have been, at the very least, twelve and a half *million* plant cells in each cubic foot of water.

Such figures are always for particular places and particular times. I doubt whether we know enough to give averages for the sea that would have much meaning. It is clear that the seas, like the lands, vary greatly in their productivity; and that the controlling factors in the sea, as in many place on land, may be the availability of a few important chemical elements, especially phosphorus and nitrogen. The teeming microscopic plants on the surface seas may use up these elements locally or seasonally, so that continuing growth depends on materials washing in from the land or welling up from the depths of the sea. Thus, for chemical reasons, some areas, like the Grand Banks or the North Sea, may be particularly rich; others, particularly poor. The Sargasso Sea, for this reason, is one of the least productive parts of the oceans. Despite the mats of floating weeds, the water is extremely clear, with relatively little plankton growth. No part of the sea, however, approaches the sterility of the great land deserts.

The microscopic plants that form the basis of the economy of the sea depend, of course, on light as the source of energy that they use in synthesizing starches. The depth to which light penetrates in sea water depends on several factors: on the angle at which the lights hits the water surface, on surface conditions—whether smooth or broken by waves—and on the transparency of the water itself. Within the water, the longer waves are absorbed most rapidly, which means that the reds and yellows disappear first and that the blues and violets penetrate to the greatest depths. William Beebe, diving in his bathysphere off Bermuda, could still see light at 1700 feet, but it was completely blue light. Traces of light can still be detected at 3000 feet in the open ocean in the tropics by means of photographic plates, but at greater depths all light disappears. Photosynthesis depends chiefly on light of relatively long wave length, so that the phytoplankton, despite this great penetration of blue light, is limited to the upper 200 feet or so of water. All life below must depend on organic materials that drop down from this surface zone.

For a long time it was thought that life could not exist in the eternally dark and cold ocean depths because of the enormous pressure of the water. The first clear evidence that this was wrong came in 1858, when one of the recently laid marine cables in the Mediterranean broke and was hauled up for repair. It was found to be encrusted with bottom-living animals which had clearly grown at depths as great as 1000 fathoms. The fallacy of the pressure argument then became clear: water pressure in itself would have no ill effect on living protoplasm, provided the organism had no spaces filled with air or some other gas. The pressure would be exerted equally on all sides and would be the same inside and outside the organism. After all, we live under an air pressure of 15 pounds to the square inch, but we cannot feel it and can only detect it by creating a vacuum. We have trouble with water pressure because we carry air into the water with us.

In the years after 1858 great progress was made in devising apparatus for collecting samples of bottom deposits and of water at great depths. The deep ocean bottom was found to be covered everywhere with a fine ooze made up chiefly of the skeletons of microscopic protozoans. The bottom ooze was found also to contain a strange, gelatinous substance, which looked like undifferentiated protoplasm. Thomas Henry Huxley decided that this must be the most primitive of living stuff, the very beginnings of life, and he named it *Bathybius haeckelii* after the great German biologist, Ernst Haeckel. It was presently found that this *Bathybius* was an artifact: the gelatinous stuff resulted from the action of alcohol on the bottom ooze, and was not living at all.

But exploration of the depths continued, and a host of strange animals was found. Somehow the idea has persisted that very primitive creatures left over from the remote geological past may still live there; this seemed to be confirmed in 1938, when L. B. Smith announced the discovery in South African waters of a coelacanth, a type of fish thought to have become extinct in the Eocene, seventy million years ago. The first thought was that this survivor from the remote past must have strayed up from the depths to be caught in the fisherman's net. But the coelacanth, it is now clear, is not a creature of the depths, but of the surface waters of the Madagascar area, and had remained undiscovered for so long simply because these waters were inadequately explored. I think it is a mistake to regard the ocean depths as a possible refuge for primitive forms of life that have been unable to meet the competition of the more progressive surface world. The ocean depths form a very special environment and require very special adaptations if organisms are to survive and flourish there. It is a strange and wonderful world in its own right, and by far the most difficult part of the biosphere for man to explore.

Man, without any special apparatus, is limited to the top hundred feet of water, and this limit is achieved only by people with special training and skill, like the Polynesian pearl divers. Since ancient times, men have been trying to devise apparatus to enable them to go deeper and stay longer, but not very successfully until the perfection of the diving helmet and air pump in the Nineteenth Century. With this, men could stay under water for prolonged periods, but they still could not go very deep, the limit being somewhat less than 300 feet. Cousteau's invention of the aqualung gave the diver freedom from cumbersome armor and from dependence on surface crews, but did not significantly increase depth penetration.

Experiments with submarine ships have been going on for a long time, and in the present century such ships have attained considerable efficiency—but for purposes of war, not of natural-history observation. William Beebe and Otis Barton in 1935 built their bathysphere for deep-sea observation. This was a heavy steel sphere with thick quartz windows, built to withstand the great pressures of the deep, and equipped with compressed oxygen and chemicals to absorb excess carbon dioxide and moisture. It was lowered into the sea by winch and connected, for communication, with the mother ship above by a telephone cable. With this contrivance, Beebe was able to descend to about 3000 feet in waters off Bermuda and to report directly, for the first time, on what life in the sea at such depths looks like.

But the bathysphere was handled, necessarily, by a cable from the mother ship above, which proved to be a limiting factor. Descents of more than 3000 feet were never made because of the danger of fouling the cable—an ever present danger which would have left the occupant of the bathysphere helplessly trapped below. The next breakthrough was made by Auguste Piccard, who designed in 1958 a *bathyscaphe,* an apparatus that could descend to great depths and come up again under its own power. This is achieved, essentially, by attaching Beebe's bathysphere as a gondola beneath a sort of "blimp" which achieves buoyancy through the use of gasoline, and ballast through several tons of iron filings. Electric motors make possible a limited amount of horizontal movement. Although at the time of writing descents to more than 15,000 feet have not been reported, it seems likely that with apparatus involving this general principle, man can reach the greatest depths of the oceans.

Although man's direct observation of the ocean depths is still extremely limited, he has been ingenious in devising methods for indirect observation—for dredging samples from the bottom, for netting organisms at different depths, for making measurements of such factors as temperature, salinity and light. The development of echo-location techniques (sonar) during the Second World War proved, incidentally, to be a great help to oceanography. Rapid progress is being made in mapping the topography of the ocean floors by echo-location, and details of the submarine mountains and troughs of the Atlantic and Pacific basins are being worked out.

Submarine work with sonar devices first made man really aware of the world of underwater sound. The sea in general is as noisy as the land—despite Cousteau's title for his fascinating book, *The Silent World*—but the air-water interface acts as a sound barrier, so that noises from the one world hardly penetrate the other. The submarine listeners of the war, when they started systematic observations, came across a whole variety of sounds that had nothing to do with enemy ships. To identify all of these mysterious beeps, groans and croaks, the United States Navy Office of Naval Research started underwater sound studies that are still being actively carried out.

Whales and porpoises, as might be expected of social mammals, are very garrulous. But so are many fish; and many different kinds of invertebrates also make noises. Sound actually travels faster and more easily in water than in air, so it is not surprising that animals in the sea should have come to use sound for communication and perhaps even, like bats in the air, for navigation. The exact speed of sound in water, as in the air, varies with factors like temperature, density, salinity and the like; but roughly sound travels five times as fast in water as in air. The speed of sound in normal surface sea water, at 0° C, is 1543 meters per second, and in air at the same temperature, 332 meters per second.

Depth measurements are made by noting how long it takes a sound sent from the surface to bounce back from the ocean floor. In making such measurements, it was early discovered that there were preliminary echoes before the main echo from the bottom came back: the sound hit something in between, which came to be known as the "deep scattering layer." This deep scattering layer varied considerably in depth from place to place, and from time to time at the same place. It seems most likely now that the scattering of the sound is caused by a concentrated layer of plankton organisms several hundred feet below the surface, a layer that goes deeper during the day, and comes closer to the surface at night. Since this scattering layer may be at depths as great as 2500 feet, it indicates a greater concentration of organisms in deep water than had previously been supposed. But Beebe and Barton with their bathysphere, and the French with the bathyscaph, report seeing more life at these depths than had been supposed to exist, though they have not observed any clear-cut layer that would correspond with the sound tracings on the depth charts from echo-location.

The number of kinds of organisms, and the number of individuals, decreases as the depth increases because everything below the lighted surface areas must depend on food drifting down from above, and the greater the depth the more depleted the food supply. With increasing depth, there are also changes in the general character of the organisms, though since these changes are both gradual and irregular, it is difficult to give any precise characterization to a series of depth zones.

The peculiar deep-sea creatures are mostly 1500 feet or more below the surface during the daytime—they may come nearer the surface at night, or in certain situations where the cold waters of the depths well up. The depth distribution, apparently, is controlled by light and temperature, as well as by water pressure.

The fish at 1500 feet and greater depths tend to have large, very sensitive eyes, presumably much more efficient in perceiving traces of light than the human eye. The fish are mostly deep, velvety black or very dark, and the invertebrates tend to be dark red. Red, at depths where light of red wave lengths never penetrates, would give the same visual effect as black.

In the eternal night of the depths of the sea, then, animals tend to become either black or dark red, and to develop increasingly sensitive eyes, though some species, especially of great depths, tend to become blind. In the eternal night of caves and underground streams, animals—fish or invertebrates—tend to become white and blind. This is associated with something else. In the depths of the ocean there is an extraordinary development of animal light, of bioluminescence. But in the blackness of caves there is no animal light, no bioluminescence.

This curious difference between cave animals and deep-sea animals may be a biological accident. Cave animals may tend to be white and blind simply because no luminescent animal (except a New Zealand glowworm) ever got started on the path of cave evolution. Most permanently underground animals are aquatic, and as far as I know, there is no luminescence, no phosphorescence, in any fresh-water animals. But the surface waters of the seas have many kinds of luminescent organisms; and as surface organisms evolved special adaptations for living in the depths, one can see why surface animal light would be retained, or greatly developed. Once the light was carried into the depths, the big eyes for perceiving it and the dark colors for avoiding being seen, would follow. One successful species of luminescent cave fish might change the whole tendency of cave evolution in the region where it occurred. This, of course, doesn't explain why luminescence is common among marine animals, sporadic (fire-flies and a few other things) among land animals, and absent among fresh-water animals.

Luminescence among deep-sea animals apparently serves

several different functions. In some cases, it is just a lure for prey. The angler, a fish with a light dangling in front of its mouth, is the classic example of this. Light is attractive for many marine animals, just as it is for many nocturnal insects on land: a light hung into the water will rapidly attract all kinds of creatures. Why natural selection hasn't eliminated this reaction in deep-sea organisms is another question, but the many kinds of anglers are a proof that it still works.

Light also serves for escape or as a deterrent. There are deep-sea shrimp and at least one squid that when alarmed emit clouds of luminescent secretion. The theory is that the animals escape in the consequent confusion, as do surface squid with their ink clouds. And some of the fish, with powerful lights, may manage to scare their enemies when they suddenly light up.

The luminescence also clearly serves for recognition. The patterns of light organs on the different fish species are endlessly diversified, and the light patterns, like the color patterns of the daytime reef fish, would enable the two sexes of a given species to recognize each other, or groups of individuals to form schools. Apparently, in some cases, the lights also serve to illuminate the field of vision. This, at least, seems the most likely explanation for the species that have large light organs just before each eye.

Most of the deep-sea fish are small, minnow-sized things, though with fantastic shapes. They tend to have relatively enormous mouths and teeth, and some of them are clearly able to swallow animals larger than they are. Food is scarce in the depths, and tends to be scarcer the greater the depth, so that anything that can be got must be utilized. All deep-sea animals tend to have fragile skeletons, but strong skeletons are hardly needed and would be difficult to achieve since the calcium supply diminishes with the depths. Someone has suggested that all of these deep-sea animals have congenital rickets because, in the absence of sunlight, there is a lack of vitamin D. Certainly in these still waters fantastically attenuated forms can develop: crabs with long slender legs, shrimp with long, delicate antennae, and all sorts of feelers and projections and elongated fins on fishes. Most of the fish are small, but there are giants in the depths among some of the invertebrates—sponges, coelenterates, tunicates and such sessile animals—because a giant size is more easily achieved in this very still and uniform environment.

The biggest invertebrates of all—the giant squid—are probably not inhabitants of the great depths, although remarkably little is known about their habits. In fact, about all that is known about these squid is from specimens that have been cast ashore, mostly in Newfoundland, but also in the British Isles, Scandinavia and occasionally elsewhere. Several specimens have had a total length of more than fifty feet, but most of the length is from two very long tentacles, the body itself being only about a quarter of the total length.

The squid, cuttlefish and octopuses, collectively known as the cephalopods, are an extraordinary group of animals. They are molluscs—that is, they belong to the phylum that includes snails, clams and the like. But, as Alister Hardy has remarked, they are very glorified snails. They have developed eyes that are superficially very similar to vertebrate eyes, though with a basically different structural plan that is in some ways more efficient than the vertebrate eye; they have highly developed nervous systems; and, with jet propulsion, they have developed a method of locomotion that has both high speed and great maneuverability.

It is becoming clear that many kinds of small and medium-sized squid are very common in the open ocean; but because of their agility they are not easily caught in nets or traps, and since they come to the surface mostly at night, they are not easily observed. Most of what we know about squid and other cephalopods concerns species that live near shore—species that are more readily observed and that can be kept in aquariums.

But the life of the open ocean is in general difficult to study, requiring special oceanographic ships and all sorts of complicated apparatus and skills. On the other hand, the life of the seashore is open to anyone who can get to the coast, and any coast holds wonders that can easily absorb a man for a lifetime. Instead of trying to describe the life of the margins of the seas in general terms, it seems to me that it would be more interesting to look at a particular environment—the coral reef—in some detail.

5. The Coral Reef

Full fathom five thy father lies;
  Of his bones are coral made;
Those are pearls that were his eyes:
  Nothing of him that doth fade
But doth suffer a sea-change
Into something rich and strange.
Sea-nymphs hourly ring his knell;
                    Ding-dong.

—SHAKESPEARE, in *The Tempest*

Coral reefs are essentially tropical. The reef-building organisms will not grow in waters where the temperature falls much below 70° F. Growing reefs are found outside the tropics in a few places—Bermuda, for instance—but in all such places warm currents bring tropical waters into higher latitudes. Within the tropics, coral reefs are rare on the Pacific Coast of America and the Atlantic coast of Africa, because of cold currents. The richest reef developments are in the Western Pacific, the Indian Ocean and the Caribbean; most fabulous of all is the Great Barrier Reef of Australia. Many kinds of corals, like the precious coral of the Mediterranean, grow in colder water, but they do not form reefs.

With the wide availability of watertight goggles or masks, and other apparatus for skin diving, the world of the coral reef has been opened up for easy exploration by man. This is true, of course, not only of coral reefs, but of all habitats near the surface in clear water. We can see the development of a new and very fascinating natural-history occupation—fish watching. People generally like to look at fish; this is clear enough from the crowds that visit an aquarium and from the popularity of fish bowls or fish tanks in the living room. But with goggles or masks, man's relations with fish have taken a new turn. Instead of putting the glass around the fish, the glass is put over the observer's eyes. The fish and the observer are thus both free to go their own ways and follow their own impulses.

Comparison between fish-watching and bird-watching is inevitable. Equipment in either case ranges from the very simple to the complex. Mask and flippers represent the minimum for fish watching; binoculars for bird watching. Complexities include things like telephoto lenses for photographing birds, waterproof camera cases for photographing fish; blinds for birds, aqualungs for fish; and so forth. The more one knows about the fish or the birds, the more fascinating the watching can become, though a deal of pleasure can be got with a minimum of information in either case. There is one big difference: anyone can watch birds in his own back yard, though New Yorkers may not be able to watch anything much except pigeons. Only the people who happen to live in some such place as the Florida Keys can watch fish in the back yard. But there is an advantage with the fish that makes up for this: the fish, for the most part, don't mind being watched, while the birds, from longer and more intimate experience with man, have become understandably suspicious and tend to be nervous when aware that they are being watched. Those peeping into the home life of birds must resort to stealth or build special hiding places. With fish it is only necessary to be relatively quiet, patient and well mannered.

Man's eyes are of little use under water, and fish watching depends on keeping the eyes in a pocket of air enclosed by some transparent material—goggles, masks or helmets. It is curious how long it took Western man to realize this. Helmets have been around for some time, but the complications of helmet diving gear restricted its use to professionals. The Japanese have long known the advantages of glass goggles, and the Polynesians saw the possibilities when they realized there was such a thing as glass. But only in the 1930's did this idea spread from the Pacific to North America and Europe. The addition of an air tube, or snorkel, to the mask is even more recent. Many of us, as children, must have played around with the idea of using a rubber breathing tube to stay under water, but no one turned this into a practical apparatus until recently. Now, at any seaside resort, masks and snorkels are everywhere, and everywhere you see a few free spirits heading out to sea with that remarkable invention of Cousteau's, the aqualung, or with some modification of his basic idea. Flippers came along incidentally, because only when you could see did it become important to have the hands free—for spearing, for photographing or for taking notes on a plastic slate.

There are plenty of hardy souls who go in for skin diving or fish watching in cold northern waters, either fresh or salt. But the full possibilities of this way of life can be realized only in clear, tropical waters, and above all, in the waters around a coral reef. Coral reefs have not always been easy to get to; but judging by the proliferating hotels of the Caribbean and the tropical Pacific, and by the swarms of tourists who descend on them, this is no longer true. And anyone who has gone to the vicinity of a tropical reef and not watched the fish there has missed one of the most interesting experiences available to man.

The reef world is a topsy-turvy sort of place from the point of view of intruding man, used to the arrangement of living things on land. The basic principles of biological organization are the same in the reef situation as in any biological community, but the various roles in this organization are played by quite different sorts of organisms in a reef as compared with, say, a forest. We are used to seeing the fixed, organic structure of the landscape built up by the green plants—by seed plants, conifers and ferns; by grass, shrubs and trees. But these so-called higher plants which dominate our landscapes are, with only a very few exceptions, absent from the sea. The fixed structure, the living scenery, of the reef is mostly made by animals—plant-like animals that cannot move around. The basic columnar trunk and branching form of trees and shrubs is replaced by a bewildering diversity of shapes. Instead of the general green of foliage, varying only in shade, the scenery is built up from the whole spectrum—blues, greens, browns, reds, purples, mostly in pastel shades. Even in the attached algae, which are true plants, the green of chlorophyl is generally masked by other pigments so that the plants are red, brown or purple.

The corals themselves are animals—tiny polyps ranging from the size of a pinhead to that of a pea, according to the species, living as immense colonies in the limestone skeletons that they have jointly built. The shape and sculpturing of the skeleton depends on the particular species of coral. Some species form smoothly rounded masses, with a surface molded like that of the human brain—hence they are called brain corals. Others form branches of various sizes and shapes, like fingers or, in the staghorn corals, like the antlers of deer. The coral animals are nocturnal, extending their tentacles to filter their food from the microscopic life of the water only at night. But they have, living at the surfaces of their bodies, microscopic plants, green algae, which are busy storing up energy from the daytime sun.

The sea fans, waving their fronds gently in the currents, are also colonial animals—close relatives of the corals, but forming their skeleton mostly with a hornlike substance rather than with rigid limestone. And there is a host of other kinds of fixed, immobile animals: sponges, generally with no resemblance to the sponges we know in the bathroom, sea anemones, tunicates, barnacles, molluscs and the like.

This animal world, like the animal world everywhere, is dependent ultimately on plants, but the plants are different from the ones we are used to. The fixed animals can depend on their food—plant or animal—being brought to them by the water currents; and the basic plant life of the reef or of any marine situation is the invisible alga multitude in the floating plankton. Many of the fixed animals, like the corals, have some of these algae living right with them, in the relation that the biologists call symbiosis.

There are, of course, a few plants that look something like plants ought to look—various kinds of green, brown and red algae. And part of the mass of limestone growth is made by plants, the coralline algae, which look something like gigantic lichens in their delicate branching form, though they are red or purple instead of green. These coralline algae are numerous enough, often, to contribute greatly to the growth of the reef.

The action, before this weird backdrop, is carried out almost exclusively by gaudy fish. Occasionally, to be sure, a jellyfish will pulsate slowly across the stage, in its ineffectual fashion, trailing its tentacles behind. Occasionally a small detachment of squid, practicing precision maneuvers, will pause long enough to assess the situation; or rarely an octopus, mistakenly nervous about being detected, will dash across the field of vision and settle down in full sight on some new coral head—to disappear completely. But the dominant movement, the action, the life of the reef is provided by the fish. They play all the visual roles that on land are divided among the birds, butterflies, squirrels, lizards. But this gives no feeling of sameness or monotony. You have to stop and think about it to realize how fully the daytime action on the reef is monopolized by this single group of animals.

A single group of animals, yes, but endless in their variety. The names show this: butterfly fish, angelfish, damsel fish, squirrel fish, parrot fish, triggerfish; as well as wrasse, groupers, grunts, pipefish, snake eels, scorpion fish. There will be hundreds of kinds of fish around a single reef area: clouds of metallic blue or green wrasse; angelfish wandering over the corals with slow dignity; big-eyed squirrel fish suspended in midwater, staring with unwavering intensity at some spot of coral. Damsel fish—perhaps more properly called demoiselles—will be scattered singly at discrete intervals over the reef, each little fish in the territory that it has staked out as its own and that it defends fiercely from any intruder, belying its gentle name.

The overwhelming impression, whether of fish or of background, is color: exuberant, varied, striking color. Inanimate nature sometimes shows a comparable abandon of color, as in the Painted Desert of Arizona, but nowhere else in the living world does color reach the variety and dominance that it shows in the coral reef. Why? No one really knows the answer—perhaps there is no single answer.

The green of landscapes is due to the chemical nature of chlorophyl, just as the red of blood is due to the chemical nature of hemoglobin. The colors are incidental or accidental consequences of the chemical structure of the molecules of substances that play important parts in the chemistry of living. A variety of color may thus be an incidental consequence when chemical processes alone have free sway. Ordinarily the free play of color resulting from these chemical processes is masked and controlled by biological processes. The green of the chlorophyl is chemical, and chlorophyl, the basic substance through which plants build up starch from carbon dioxide by capturing energy from the sun, dominates the living world on land. The seed plants, with sporadic exceptions, have not masked this green of chlorophyl with other pigments. Insects, birds, lizards, a host of animals living among foliage, then also become green—not as a consequence of their chemistry, but because green serves a biological purpose, to make them inconspicuous either to their enemies or to their prey. The dead tissue of the seed plants turns brown or gray, and we also find a host of animals living on tree trunks or in the litter on the ground that, for biological reasons, have taken on the browns and grays of their surroundings.

The bright and varied colors and patterns of flowers serve as signals to attract the insects necessary for their pollination. Bright colors are sometimes warnings—the bright wasp advertising its sting, or the gaudy butterfly notifying predators that it has a nasty taste. We find, among animals, all of the principles of camouflage, of counter-shading, and broken silhouettes. We find, sometimes, bright marks that seem to serve chiefly for recognition, so that males can find females or vice versa. Sex often determines coloration, and Darwin long ago explained peacocks, pheasants and the like in terms of females picking pleasing mates.

Ordinarily, then, we can explain the colors of the living world in terms of chemical processes masked or modified by the exigencies of life, by the needs of catching food, or escaping from becoming food, or finding mates. Our explanations are sometimes dubious, sometimes far-fetched. Often there are several alternative theories to explain a particular situation. The whole thing at least seems fairly reasonable, but this doesn't help much when we put on our masks and start looking at the colors of the coral reef.

There are plenty of beautiful examples of protective coloration, like the pipefish that stands on its nose to become one more blade of waving sea grass; or the octopus that instantly changes its spots to fit the particular chunk of coral on which it has settled. There are also many examples of warning coloration, of creatures that advertise their deadly stings or poisonous flesh. But this doesn't explain the pastel blues and pinks and purples and browns of the corals and algae, or the startling metallic brightness of the swarms of tiny wrasse, or any of the other flauntingly conspicuous but perfectly edible and harmless fish.

One is sometimes tempted to think that here on the reef chemical processes have gone wild, lost the biological controls that govern the rest of the living world. Life has responded to the bright sun, the crystal water, the supporting warmth, with an irrepressible exuberance in form and color. This theory gains plausibility when, on a Pacific atoll, you have opened a parrot fish just roasted over coals from coconut husks, and find, not prosaic fish bones, but delicate bits of jade green embedded in the white flesh. What possible meaning can this have in terms of the struggle for existence?

Much of the background color of the reef, the color of the corals, algae, sponges and other fixed organisms, is probably of this physiological sort. In the case of algae, the red and purple pigments enable the plants under water to absorb light more efficiently for their photosynthetic purposes. In the case of corals, the colors are in part due to pigments laid down by the coral animals themselves, and in part due to microscopic algae living symbiotically with the animals. It is hard to imagine what function some of the coral colors serve: there is a lot we need to learn yet about these corals.

In the case of the active animals of the reef, it seems likely that most of the colors have ecological, rather than physiological, meaning. The life of the reef abounds in cases of concealing coloration, particularly among the crabs, shrimp and some of the fish. These animals, caught and taken out of the water, often look extremely bizarre, but their bizarre colors and forms match perfectly the irregular and colorful environment in which they live—which is why they are hardly noticed by the diver. But a large proportion of the fish plainly are advertising their presence, making no attempt at concealment. The commonest explanation of this is that the colors and patterns enable the fish of different kinds to recognize each other. It is, incidentally, also a great help to the skin-diver who wishes to recognize the different kinds of fish. But why, then, don't all fish have such striking recognition colors? It is surely just as important for fish to recognize each other in the North Sea as it is in the Caribbean.

One answer to this is that there are far more kinds of fish in a tropical-reef situation than there are in other marine environments. Another is that in very clear, sunny, tropical waters, visual stimuli of all sorts take on particular importance. Where light is less reliable, other senses like smell and hearing must take over. The fish with bright colors and contrasting patterns are mostly daytime animals. At night they disappear into their hiding places in the coral, and the reef is taken over by a different set of animals.

Fish-watching at night, of course, presents some special problems, mostly turning on differences between man and fish. Man is primarily a daytime land animal. With mask and flippers, he makes the transition into the water fairly easily during the daytime, when he can depend on sight to tell him what is going on. The fish around him also depend on sight, but in addition they have a keen sense of smell, and a sort of sixth sense in their lateral-line organs that is difficult for us to appreciate. It is as if, with eyes shut and ears plugged, we were able to sense exactly the position of a bee buzzing around our head. The development of such a sense depended on the nature of water, where movement causes a more appreciable disturbance than in air. Man, with his goggles, has a real advantage over fish during the daytime, despite his lack of smell and lateral-line sense, because of his eyes. The big difference between the human eye and the fish eye is in distance perception. On land, we can see for miles; in the water, if it is really clear, we can see as far as a hundred feet. No fish can match this, since fish eyes are built for close-up vision. At night man has lost this advantage.

To be sure, it is simple enough to arrange for waterproof lights. There you are in a sea of blackness, with a narrow cone of light that under the best circumstances penetrates the gloom for only a few feet. You know that the big cats of the water—the sharks, the barracuda, the moray eels—are most active at night. You know a great deal about their daytime habits, but how can you, from this, be sure about what they will do at night? As a result, you spend your time turning around and around to be sure that nothing has sneaked up behind you—it's a scary feeling.

It is more comfortable to watch reefs at night from the security of a glass-bottomed boat. The stage is unchanged, but at first it looks strangely empty because most of the gaudy fish are gone. They are in hiding for the night. Each kind has its distinctive sleeping habits. Some kinds of parrot fish even secrete a big glob of mucus in which they sleep, apparently quite secure (laboratory experiments have shown that fish in their envelopes are protected from the attacks of moray eels).

There are lots of moray eels on reefs, though you don't see much of them during the day. Morays are among the most vicious creatures nature has devised. They look it, too. If you watch a patch of reef closely, you may see a snakelike head sticking out of a hole, slowly weaving back and forth like a cobra under the spell of a charmer. The moray is scannning the water around him for the smell of some possible victim, mouth wide open, showing the multiple rows of fanglike teeth. Moray eels have never been known to make an unprovoked attack on man, and during the day they usually stay well withdrawn into their holes. Every fish-watcher knows, or ought to know, that you never stick your hand into any part of a reef when you can't see what is there, because any hole may have a moray in it, and the moray may resent the intrusion. When a moray bites, he never lets go—you have to cut his head off. Morays don't see very well, and so it is not surprising to find them gliding about in the open at night, searching for their prey, when they can rely on their incredibly keen sense of smell. Morays come in assorted sizes and colors, the biggest reaching a length of about seven feet. They are fascinating to watch from the safety of the boat.

The squirrel fish also carry over from the day scene to the night. There are many kinds of these also—but all of them are some shade of red, and all have big eyes, which in animals is frequently a sign of nighttime or twilight activity. Occasionally you see other fish, but it is often difficult to be sure whether they are really nocturnal, or whether they have been awakened by your light.

Sometimes you see queer things on the reef at night. The big spiny sea urchins, for instance, which you think of as always half-hidden, each quietly resting in his little pocket among the corals, go promenading. You can watch them on a patch of clear sand, where there were no urchins during the day, in slow-motion parade. In the magic of the night the wooden soldiers have come to life, though it is a stiff, hardly perceptible life.

But fish-watching is essentially a daytime occupation, and it can really become rewarding when followed systematically for the purpose of finding out about fish habits. One of my fish-watching colleagues, John Bardach, has spent quite a lot of time concentrating on parrot fish. Something over a hundred different kinds of parrot fish have been described from the Pacific and the Caribbean. They get their name not because of their color—though the fish and the birds are about equally gaudy—but because of their parrot-like beak. This serves them well for browsing and scraping among the corals and algae. Those most frequently seen are a foot or so in length, usually moving about in small schools; some species attain a considerable size, weighing a hundred pounds or more. It is hard to generalize about their color, but greens and blues predominate among the gaudiest ones, some are red, and still others have black and white patterns.

Watching schools of these fish, Bardach noticed that not all members of a group looked alike. One fish would usually be brightly colored, the others plainer. They would slowly move across a reef area, grazing. He saw that the brightly colored member of the group would sometimes show peculiar behavior, dashing out as though to pick a fight when the group happened across a solitary fish of the same kind. After seeing this a number of times he began to wonder—the bright fellow was acting like a stag or bull protecting his harem. You can recognize a stag or a bull, but how do you tell the sex of a fish? Actually, with most fish, there is no way of telling except by cutting them open. He did this—and found that the bright fellows were always males. This was surprising, because the bright and plain fish that schooled together had been described as separate species.

Clearly it was important to find out more about this, so Bardach tried experimenting with the fish in aquariums. He took some of the plainer parrot fish and injected them with male sex hormones. Sure enough, they took on the bright male coloration. He found this to be true of a variety of different kinds of parrot fish—so that it began to look as though there might be far fewer different kinds of parrot fish in the world than had previously been supposed.

The parrot fish, with their constant browsing, play an important role in the breakdown of the reefs. They not only nibble on algae, but also scrape off considerable quantities of lime material, apparently on purpose. If you listen, you can hear a characteristic crunching noise made by these bony beaks—it doesn't take much imagination for your own teeth to start aching. I am not sure what purpose all this limestone serves for the parrot fish. It may serve to mince the food, like the grit in a bird's gizzard, or it may serve some chemical digestive function. But whatever its purpose, the limestone itself is indigestible and what went in must come out. The parrot fish consequently, regularly defecate clouds of fine, white sand.

Parrot fish, like other browsing animals, feed almost constantly during the day. Bardach, in the course of his parrot-fish studies, got curious about this sand business and tried to make some rough guesses as to the quantities that might be involved. First he had to find out how long it took materials to pass through the gut of the fish. He did this by feeding some fish on charcoal, and timing its reappearance. He found that this varies with the size of the fish—taking from two to eighteen hours. He repeatedly counted the number of parrot fish on various reef areas around Bermuda, and found this to be quite variable, but averaging somewhere around a hundred individuals per acre. He also weighed the sand in the gut of a number of fish. By some admittedly extremely dubious calculations, he guessed that the parrot fish deposit at least a ton of sand per acre on these reefs every year. If you consider that the Bermuda reefs occupy several hundred square miles and have existed in their present shape for several thousand years, the parrot fish account for quite a lot of sand.

Other browsers on the reef are the surgeonfish, quite similar to the parrot fish in their feeding habits, but looking very different. They are called surgeonfish because of a pair of stiletto-like folding spines on their tails, which are supposed to serve some defense purpose, though I have never seen them used. Many Pacific species are very brightly colored, but the Caribbean ones are less spectacular—generally blue or brown, though the young of one common species are bright yellow.

By browsers I mean herbivores—the fish that live chiefly on algae. The butterfly fish are very similar, but they might be called nibblers because they chiefly select tasty animal morsels from among the rich growth of the reef. They are among the most bizzare reef residents in shape and coloration. Like most fish that don't swim great distances, they are thin and high rather than cigar-shaped, and extremely adept at dodging. Most have striking patterns of bars, stripes or spots. The four-eyed butterfly fish, for instance, has a prominent black eyelike spot above the base of its tail, which is generally supposed to serve to fool predators into thinking that the tail is the head—fish predators generally catch their prey head first, since fish cannot swim backward to escape. Butterfly fish are generally found in pairs, and on examination these are found to be male and female. It thus seems probable that these are unusual among fish in being paired off for longer than just the spawning period.

The trunkfish are another group of nibblers. The first time you see a trunkfish, you are struck by its unlikely appearance. It has a tiny pursed mouth with sharp razor-like teeth, then a rigid body, triangular in cross section, out of which sticks a seemingly ineffectual rudder-like tail. You may think: surely this is a fish I can outswim; and the trunkfish will egg you on into believing this. You follow it and it stays just a few feet ahead of you; you almost think it will look around to see whether you'll take up the chase, but every time you think you have gained a foot or so, it gives one or two strokes with its tail and you're just where you started. Finally you lose it, with the realization that after all you are merely an awkward intruder into this strange medium where the fish are at home.

The groupers, among the reef fish, look much more ordinary, although some of them have quite striking patterns. The Nassau grouper, for instance, a black-and-white-striped fellow, is called the chameleon of the sea—by contraction and expansion of its black color cells it can change its shade and pattern almost instantaneously to blend in with any background from nearly creamy white to nearly jet black. Some groupers get to be quite big, and though they are very tame and nonaggressive, it gives you a somewhat eerie feeling to face a large grouper in its hole sitting there quietly watching you with his mouth half open, realizing that he has quite sharp teeth and that he can easily swallow things as large as your head. But while groupers look fierce enough, they can be turned into pets by being offered food. But so can lions.

It seems that large groupers have a feeding circuit and may visit several different reef areas on their tours. It is thus difficult to decide whether to call them residents or visitors on the reef. But some of the fish that you see are clearly visitors rather than residents; among these are the sharks and the barracuda. Divers have been busy for some time now deflating the reputation of sharks, and the sharks in Pacific lagoons or around shallow Caribbean reefs are generally not dangerous kinds, though one tends always to treat them respectfully.

There was a nice demonstration of this in the case of a student fresh from the tame lakes in Michigan, who on his first day on a Bermuda reef aroused the curiosity of a six-foot shark. Usually sharks stay at a respectful distance, for after all you are nearly as big as they are, but this one apparently liked people, because he came right up to the diver, whose only recourse was to bat him over the nose with the plastic slate used for note taking. The shark took the hint and went off.

With barracuda you can't use this tactic because they don't come close enough. Barracuda and sharks have quite different feeding behavior. Most sharks have rather poor vision and rely a great deal on smell—hence the caution against trailing speared fish around with you in the water. Barracuda have much better sight and are known to strike at flashing objects. Of course, fully documented accounts of attacks by either sharks or barracuda are rare, but it is possible that a barracuda would strike at a buckle or bracelet or a trailing speared fish. I have never heard of barracuda attacking skin-divers—they are more apt to go for painted toenails dragging from a boat—but they are common around reefs and look nasty enough, with their sleek bodies and their habit of swimming with mouth half open, showing their formidable teeth. Usually they seem to pay no attention to the diver, but sometimes they circle persistently. You can usually call their bluff by lunging at them.

Morays, sharks, barracuda—these are chiefly psychological

hazards. Trivial hazards, the nettles and poison ivy of the reef, are more commonly encountered. Probably the greatest hazard of the reef is sunburn. Then there is the coral itself, which is rough and cuts easily, and coral cuts are notorious because they heal so slowly, especially if small fragments get imbedded in the wound. There is a special group of coral relatives, the fire corals, which cause a nettle-like sting if you brush against them. One soon learns to recognize their smooth-appearing surface; color and shape are variable, but they tend to be flesh-colored or orange, and they often grow upright, with finger-like or blade-like projections.

There are quite a few other things that may sting or burn, like jellyfish and Portuguese men-of-war. The ever present sea urchins are prominent members of the nuisance class, although some of them with poisonous spines can cause a pain that goes beyond nuisance intensity. But there are only a few things that are really dangerous—a Pacific cone shell that, if handled carelessly, can inflict a sting with its radula that may be fatal, and a few fish with poisonous spines, such as the lion or scorpion fish.

These hazards and nuisances are inconsequential when compared to the pleasure that can be got on the reef. With the invention of the spear gun, man has started hunting on the reef—and this, like hunting everywhere, causes changes. The hunting is more for trophies than for food, but even so it can lead to a rapid reduction of the bigger fish, while the others will soon enough learn to be shy. And as the man with the spear gun becomes prevalent, he rapidly takes first place among the hazards, not only to the fish but to his fellow men.

This is becoming recognized. In Bermuda, for instance, there is a restriction on the kind of spear you can use—the high-powered, trigger-operated gadgets are prohibited. There are plans for the new National Park in the Virgin Islands to include reef areas where similar or even more stringent restrictions will be enforced. This seems eminently sensible in view of the irrevocable changes that man can inflict on any environment. There should be places reserved for the fascinating pastime of fish-watching.

6. Lakes and Rivers

And an ingenious Spaniard says, that rivers and the inhabitants of the watery element were made for wise men to contemplate, and fools to pass by without consideration.

—IZAAK WALTON, in *The Compleat Angler*

Coral reefs are lovely places, but they are, for most of us, far away. Rivers, lakes and ponds, however, are at everyone's back door, and while the life of a pond may not be as spectacular as the life of a reef, the marvels are of the same kind. I have spent many hours of my life sitting quietly by ponds or looking into the water as my canoe drifted over the shallows of a lake, trying to understand what was going on in that world shut off by the surface of the water.

A pond, especially, has the fascination of the miniature. It is a world clearly limited by the shores and bottom and surface; a sufficiently self-contained world, a small enough world so that you should be able to figure out everything going on there, describe it, analyze it, perhaps fit the relations among the living things there into neat equations—and solving the equations, solve all mysteries. Understanding the pond, one would understand the biosphere. But the pond still eludes me.

The life of the pond is one incident in the network of the biosphere. The water of the pond is one incident in the system of the hydrosphere.

The free water over the surface of our planet forms a single, vast circulating system, with the sea as the main reservoir and the sun as the furnace providing the energy to keep the system circulating. We can distinguish easily enough between the atmosphere and the hydrosphere, but the two are completely interdependent, with gases from the atmosphere constantly entering into solution in the liquid of the hydrosphere, and with water from the hydrosphere constantly evaporating into the atmosphere. And life, as we know it, depends on this interlocking system.

The warmer the air, the more water vapor it can hold. Thus in the shortest version of the circulating system, water evaporates from the surface of the sea into the warm air above it. This warm, moisture-laden air is carried upward by atmospheric circulation and cooled until the water is precipitated out in fine droplets to make the mist that we call clouds and, on further cooling, to make discrete drops of water large enough to be pulled back to the earth by gravity as rain.

Water is thus constantly circulated from the sea to the air and back again. But the water in the air is often carried over land, with all sorts of consequences. The water that falls on land may flow directly over the surface to form streams in gullies, the streams converging into larger streams until at last they pour into the sea again. Or, more commonly, the water may sink into the soil to unite with the vast accumulation of ground water that underlies all land. The ground water emerges here and there as springs or seepages to start the flowing surface systems of streams and rivers. Often the flow of these streams is blocked by the contours of the land, to make pools, ponds and lakes, where the water is dammed up until it reaches some level at which flow can be resumed. But free water in contact with air always tends to evaporate, and sometimes the evaporation is fast enough so that the water never accumulates sufficiently to allow free flow again, and we have dead seas, lakes with no outlet, little side eddies in the great system of planetary water circulation.

Much of the ground water, of course, is picked up by the roots of the land vegetation and returned directly to the atmosphere through the process we call transpiration. Man has lately taken to tapping the ground water, thus introducing new short circuits into the system, sometimes with unfortunate results from his own point of view. Some of the ground water may be trapped in sealed-off pools, out of circulation, for long periods of geological time; some of it returns to the reservoir of the sea through underground channels without ever reaching surface circulation.

There are thus many sorts of pathways for the water interchange between hydrosphere and atmosphere, even though they all turn eventually on the evaporation-precipitation sequence. And the variety of these pathways governs the varied manifestations of life outside the seas.

Water, when it evaporates, does so as water—that is, it leaves all the accumulated dissolved materials behind. It starts out pure each time, but since water is the most universal of solvents, it never stays pure very long. Even as rain, it picks up not only gases from the atmosphere, but traces of all sorts of other things present as dust in the air. Rain water, however, is still reasonably pure, and the solution process really starts after the water hits the ground. The chemical nature of fresh water, then, always depends on the nature of the soil or rocks that it has passed through or over; and this chemical nature varies greatly in different places, with important consequences for the organisms that depend on the water. The proportions of dissolved salts in the vast and continuous reservoir of the sea is constant, but the fresh waters of the earth vary greatly. The total salt content is never as great as that of the sea except where the cycle is blocked, as in lakes without outlets. Then, as in the Great Salt Lake, the salt content may be much greater than that of the sea, and with quite different proportions among the dissolved salts.

Inland waters, then, always differ from the seas simply as a chemical environment. They also differ in many other respects that are important in understanding the differences in the development of life. The most fundamental of these is the discontinuity of inland waters both in space and in time.

The seas, as we have seen, form a continuous system over the face of the earth. To be sure, they are broken up at mid-latitudes by the continents, with the result that shore organisms, intolerant of cold water or unable to cross the open ocean, are confined to particular regions like the Atlantic coast of America, the South Pacific, the Indian Ocean, or even some particular island shore, archipelago or strip of coast. But the barriers to communication through the seas fade when compared with the barriers to communication among the different river and lake systems of the continents. There is simply no way of getting from the Amazon to the Ganges without a change of medium, and there never has been.

Fresh waters are discontinuous not only in space, but in time. The pattern of lakes in the north central United States is a consequence of the last glaciation. Just yesterday, in geological time, there were no lakes: the whole land surface was buried under a vast sheet of ice. And as the mountains have shifted, so have the patterns of the rivers. A few of the lakes of the world —notably Baikal in Asia, Tanganyika in Africa and Ochrida in the Balkans, have a respectable geological age, which has permitted the evolution of rather special forms of fresh-water life. Some of the great river systems of the world are also quite old. But the history of Lake Tanganyika or the Amazon River in particular, or of fresh waters in general, bears no comparison with the uninterrupted history of the seas.

This has several consequences. From the point of view of evolution, one can in general look at the sea as a thing in itself, reconstruct the history of its inhabitants without reference to either fresh water or land. The striking exceptions here are among the vertebrates. It is generally thought that the bony fish (as distinguished from the elasmobranchs, the sharks and rays) started their evolution in the fresh or brackish waters of Silurian estuaries several hundred million years ago; but if they started in fresh water, they soon and successfully reinvaded the sea, and they have flourished both in the sea and in fresh water ever since. The reptiles, though essentially a land group, developed many marine forms in the ancient times of their great glory, and still today are represented in the seas by a few species of sea turtles, sea snakes and crocodiles. Several groups of mammals, including the seals, manatees, porpoises and whales, have successfully adapted to the sea after a long history on land. But these, and a few other groups like the turtle grasses (a variety of seed plant) are still exceptional when compared with the mass of kinds of things living in the sea.

The history of life in fresh water, on the other hand, cannot be understood without constant reference either to the sea or to dry land. The clams, crayfish, worms and inconspicuous sponges of fresh water are separate offshoots from basically marine groups. Fresh waters teem with insects—many different kinds of insects representing quite different adaptations of land-living, air-breathing forms to the water medium. Among vertebrates, the amphibians started their evolution in fresh water and have stayed with it, while reptiles, birds and mammals shift between water and land in all sorts of different ways. The obvious vegetation of inland waters is composed of many sorts of seed plants derived from land ancestors.

The fact that so many kinds of fresh-water organisms are derived from land ancestors means that a great many of them, though living in the water, continue to depend on air for respiration. Some of the insects have learned to do without air, getting oxygen through their skin or through various sorts of gills, but most remain air breathers. Some of them, like mosquito larvae, have developed special tubes that can get air by breaking the water surface; in some mosquito larvae and in water scorpions these tubes are absurdly long, longer than the body of the insect. Other insects, like the diving beetles, capture bubbles of air which they can carry down with them into the water. Frogs, crocodiles and hippopotamuses have developed raised nostrils and raised eyes that are curiously similar among these completely unrelated animals, so that they can breathe and see while their bodies remain under water. Many plants, like the water lilies, have developed leaves that can float on the surface of the water or protrude above the water.

The numerous plants with floating leaves bring to mind the curious importance of the surface in fresh water. One can discuss the life of the sea with little reference to the surface film of the water. The organisms are divided into the drifting plankton, the swimming nekton and the bottom-living benthos. But in fresh water one cannot ignore the surface-attached neuston. These include things like the water striders and whirligig beetles that live on top of the surface film, and the various insects like mosquito larvae that live hanging down from the film. Wave action in the sea makes surface existence more hazardous than in small bodies of fresh water, though there are still some organisms in the sea that depend on the surface, like the Portuguese man-of-war with its big bladder-like sail, and the floating mats of Sargassum weed in the tropical Atlantic. I suspect that it is not the lessened hazards but the numerous air breathers of fresh water that give the surface its special importance in this environment.

Because of the discontinuity of fresh water, its inhabitants have special problems of getting about. A great many freshwater habitats are really disconnected and temporary—puddles that persist for only a few days or weeks, ponds and swamps that dry up part of the year, water that accumulates in rot holes in trees. There is every gradation in permanence, from the puddle that dries up within an hour or two after the rain stops, through glacial or volcanic lakes that may persist for a few thousands of years, to Lake Tanganyika or Baikal or the Amazon River with countless millions of years of history. The smaller and more transient the water accumulation, the more important the means of locomotion become for the inhabitants—having some means of getting there in the first place, and of getting away later.

The aquatic insects fit this requirement perfectly, and they dominate all sorts of small accumulations of water and the protected parts of larger lakes and rivers. With most aquatic insects, only the growing stages—the larvae or nymphs—live in water, and the winged adults are able easily enough to move from one water accumulation to another to lay their eggs. Where the water accumulation is seasonal, the life histories of the insect inhabitants are timed to coincide with the availability of water. For instance, mosquito larvae that live in temporary rain pools show extraordinarily rapid growth—from egg to adult in five or six days—while species that live in places where the water is permanent may take several months to accomplish the same growth.

With frogs and salamanders, as with insects, only the larval form is aquatic, except that a few species of salamanders never grow up. They become sexually mature, though retaining their larval form—a phenomenon called *neoteny*. Tadpoles, like insects, show great variation in their speed of growth, depending on the habits of each species.

The amphibians and the insects lead the truly amphibious lives ideal for transient fresh water. But all sorts of other fresh-water organisms have developed special adaptations to survive periods of drought or to get from one bit of water to another. Spores, eggs resistant to desiccation, and encysted forms are especially common. Eggs of copepods (microscopic crustaceans) and spores of algae and fungi may lie dormant in the soil for years, ready to start development when submerged in water. Such eggs and spores may be light enough to be carried by the wind as dust, or they may be transported on the feet of water birds. Some of the species of fresh-water plankton with such resistant stages have achieved cosmopolitan distribution, so that the same kind of animal is found wherever there is fresh water.

At the opposite extreme, of course, many species of fresh-water organisms are found only in one particular lake or one particular river system, showing complete isolation. In many other cases, each lake or river of a region will have a slightly different form of the same species of organism, showing that the populations are sufficiently isolated so that interchange is rare. Biologists call such partially isolated populations *sub-species*.

Species of molluscs and of fish are especially likely to have restricted distribution in fresh water. Even these, however, have developed many adaptations for surviving droughts or for getting from pond to pond. The lungfish of Australia, South Africa and South America are able to survive from one wet season to the next caked in dried mud, and other kinds of fish and some kinds of molluscs have similar capabilities. Eels are famous for their ability to slither through wet grass at night and thus get to isolated pools and wells, but many other fish, especially in the wet tropics, show a similar ability to travel over land from pool to pool.

The study of life in inland waters has been given a special name, *limnology*. The conditions of life in fresh water are sufficiently distinctive to require special knowledge and special methods for their study. This study gains importance not only because of its biological implications but also because the economy of fresh waters in many parts of the world is of practical interest to man. There are many fresh-water biological stations under both university and governmental auspices. Most of these are in Europe and North America, and consequently our ignorance of the biology of tropical rivers and lakes is astonishing; but this is only one aspect of the general neglect of biological studies in the tropics.

The limnologists make a basic division of their subject matter between running water and still water, which out of a love for Greek words, they call *lotic* and *lentic* environments. This distinction, like all such distinctions, sometimes blurs, as when a small stream runs through a marshy area, or when a big river, in its serpentine course, develops oxbow lakes. And of course there is water movement of some kind in almost all aquatic situations. But in general, the problems of life in running water and in still water are quite different.

The problem of animals in running water, essentially, is to keep from being swept away, to keep still, or to be able to make headway against the current. In completely still water, the animals must move to find food. The great development of sedentary animals like the corals in the sea is possible because the sea is always restless with waves, with tides, with all sorts of great and small currents. There are sedentary animals in lakes and ponds, but they are few and trivial compared with the sedentary animals of the sea.

If the problem of overcoming current can be solved, the water of a rushing stream is favorable for life. From the constant agitation it tends to be saturated with dissolved oxygen and carbon dioxide from the atmosphere; from its previous stage as ground water, before welling up in springs and seepages, it tends to be well supplied with the minerals essential for life. And there is organic food too, from the things that have been caught in one way or another by the current.

Plankton, in the sense that we know it in the sea and in lakes, cannot develop: drifting organisms are helpless, carried ever onward with the current, so that they cannot replace themselves. Drifting organisms are there, certainly, but they must be constantly renewed from the ponds and lakes and marshes that contribute water to the stream. They are unable to form a self-reproducing community under stream conditions. The algae, diatoms, mosses, all the resident plants of the stream, must be fixed, growing as films over boulders or attached in some way to the bed of the stream. In these films of plant life, there are many sorts of animals, relatives of forms that would drift as plankton in still waters, but here their adaptations must be for clinging, for resisting the current.

Streams, as every fisherman knows, have a rich insect life, but they are different insects from those found in still waters. The larvae of the pestiferous black flies live on the rocks of cascading streams, firmly attached by suckers supplemented by strong silk threads spun from their huge salivary glands. Other insect larvae have developed other ways of meeting the current directly, or of avoiding it by burrowing in the mud or sand or by keeping to the quiet little eddies and side pools that are found even in mountain torrents. The fish tend to be streamlined, efficient swimmers, especially those that live in the headwaters of a river system. In larger and more sluggish streams, a bottom-living fauna can develop, less subject to constant battle with the current.

Streams—and the erosion always associated with them—are major elements in shaping the physical features of our landscapes. The students of landscape form, of *geomorphology,* have developed an elaborate and picturesque vocabulary for describing the different types of streams and of stream action. They speak of braided streams and meandering streams, of beheaded streams and captured streams. Attempts of biologists to develop a classification of stream types have not been so successful. There is, to be sure, a great difference in living conditions in a tiny brook cascading over rocks high in the mountains, and in a broad, deep, sluggish river meandering across the coastal plain. But there is every kind of gradation between these extremes, so that any system of subdivisions rapidly comes to seem arbitrary.

Streams in general tend to be tied up closely with the land through which they flow—shaded through forests, open in meadows, gaining chemical elements from the soil of their beds and shores, receiving food from the plants and animals that fall into their currents. But this relationship tends to become less close as the stream grows larger, until, again, the great rivers of the coastal plains are things in themselves, understandable without reference to the physical or biological characteristics of their shores. Indeed, in this final phase, the mud of the river bed and of the shores is apt to be a creation of the river itself, deposited over millennia. The water of the river is the product of a thousand diverse brooks which have flowed over many kinds of rocks and soils, so that it too achieves a stability and independence of the surrounding world. The landscape here is subject to the whim of the river, rather than vice versa, but again the shift in emphasis is gradual and irregular.

With regard to fish, the number of species present tends to increase steadily from the headwaters to the mouth in a river system, and this can be looked at as one phase of the growing independence of the stream. This increase in numbers would hold for all truly aquatic organisms, but it would not apply to the kinds of animals that vacillate between the water and the land, like the insects and the amphibians. The lower reaches of a river even tend to develop their own plankton system, though it remains a peculiar system, since it is always drifting onward in a single direction until it reaches the estuary itself, where tides may block the flow or create countercurrents.

Rivers are the path whereby life from the sea reached fresh water, and hence eventually reached the land. The river estuaries remain a zone of transition in which sea organisms start appearing, one by one, depending on their tolerance of the lowered salinities. Many fish, like the tarpon, pass easily across the salt barrier, but only to linger briefly on the unsalted side. Some fish, notably species of sturgeon, salmon and shad, cross the estuaries and go up the rivers to spawn. The salmon, coming in from the sea and making their way against the strongest currents, leaping waterfalls, to reach again the same mountain brook in which they themselves were spawned, are the most spectacular, the most puzzling of these. How, wandering the seas, do they find their home river again, and how, in the river, do they distinguish each branch to come again to the brook from which they started? The recovery of tagged fish has shown that they do this with some regularity. The means must be a fine sensitivity to chemical differences in the water—but this is hard either to understand or to demonstrate.

The eels show the reverse habit: coming down the streams and rivers to the ocean and going eventually to the mid-Atlantic, to the Sargasso Sea, to spawn. A Danish biologist, Johannes Schmidt, was responsible for much of the careful detective work that eventually unraveled this story. The tiny, transparent eel larvae, called *leptocephali,* form part of the ocean plankton. They become bigger and bigger as they are collected nearer Europe or North America, until eventually, near the estuaries of the rivers, they are recognizable as eels, though still transparent—a stage in which they are called glass eels. Finally, making their way up the rivers, they become darker and are called elvers. Now the European and American eels differ in the number of vertebrae in their backbone, so they are distinguished as separate species. Both apparently spawn together in the Sargasso Sea area, but the *leptocephali* of each, drifting with the ocean currents, find their right continent.

I have emphasized the transitory nature, from the geological point of view, of all fresh-water accumulations. This is true, but it is much less true of river systems than of lakes. The details of the pattern of a river system shift constantly, changing even in the moment of time open to direct human observation. Channels shift, tributaries of one system are captured by another, uplift or landslides or lava flow may block a course. But the great river systems of the world, from the point of view of their inhabitants, show a considerable continuity, because these shifts for the most part are gradual; and even when river systems change radically, parts of the old system are carried over into the new.

The present river system of the midwestern United States, for instance, is new because the ancient rivers of the region were obliterated by the glaciation of the Pleistocene. But careful geological study has shown the pattern of the pre-Pleistocene river system, which persisted for a great many million years while the Appalachians were uplifted into a lofty mountain chain, eroded down, and raised again to their present form.

The great ancient river of this region has been called the Teays, after a small town where its traces were first discovered. The Teays was a mighty river, starting in the Carolinas, running north and then west, crossing through the middle of Indiana and Illinois, receiving the precursor of the Mississippi as a tributary near the present town of Lincoln, Illinois, and flowing into a Gulf of Mexico that reached far north into the continent. But in some places the modern river courses are identical with the ancient ones. The gorge of the New River in West Virginia, for instance, represents part of the course of the ancient Teays;

and the Teays and the present Ohio were identical for a stretch near Wheelersburg. There was thus ample opportunity for continuity in the river inhabitants. The pattern was not erased in some sudden catastrophe; it was changed by the slow march of geological events.

The largest of the world's rivers in terms of volume of water discharged into the sea is the Amazon. This mightiest of rivers forms a network of water channels that permeates nearly half of the continent of South America. Through the Rio Negro and the Cassiquiare Canal, the waterways of the Amazon are directly connected with the Orinoco system of northern South America. The waters of these two river systems together, though dispersed over most of a continent, can be looked at as constituting a great inland sea which has had continuity both in space and in time, and which, from its equatorial location, has probably always provided a stable and favorable environment for its inhabitants.

The result is a fabulous and special fauna, all too little known to science because so much of it is dispersed over the least explored part of our habitable globe. No one knows how many different kinds of fish there are in this river-sea, but it is something like two thousand. Among these may be the biggest fish outside the ocean, though the fresh-water record is for a sturgeon caught in the Volga River—14 feet, 2 inches long, weighing 2250 pounds. I can't find any statistics on the maximum size of Amazon or Orinoco fish, but I have vivid memories of huge fish skulls left on sandbars by fishermen who had been using dynamite: the original fish must have weighed several hundred pounds. At the other extreme are queer tiny things like the candiru, a tiny, slim catfish, notorious because it sometimes wedges itself in the urethra of the penis of swimmers, with most painful consequences for its host. And from this region too come many of the small, bright, beautiful fishes that we admire in our tropical aquariums: they are even more fascinating to watch in the clear waters of a pool in a forest stream.

Here too are electric eels and the notorious piranha—these

latter, I would suspect, the most dangerous of all fish for man, making sharks and barracuda seem harmless and cowardly. And the biggest of all snakes, the anaconda is at home here, more often in the water than out. The salt sea has contributed many inhabitants to this fresh-water sea—a special dolphin, a manatee, stingrays that nestle in the sand bars high on the tributaries where they rush out of the Andes.

Rays, piranhas, anacondas, electric eels—I make it sound a dangerous place, even without mentioning crocodiles. The upper Amazon or upper Orinoco certainly is not tame country—there are no manicured resorts. But its great fascination comes from escaping into a region of river and forest which man has so far not dominated, and the dangers are easily enough avoided with a little common sense. The biggest hazards are the hordes of mosquitoes which breed in the wet forests on the river margins, and the black flies and midges that breed in the rivers themselves. The only people I know who have passed through the Cassiquiare Canal from the Amazon to the Orinoco system report that the black flies made this unique experience a torture. But this surely could be alleviated with the new insect repellents.

Lakes can be looked at as blockages in river systems—mostly temporary on a geological time scale. The great exceptions, occupying ancient depressions in continental areas, are, as I have mentioned, Lake Baikal in Siberia, and Lake Tanganyika in Africa. A few other lakes, through the peculiarities of their inhabitants, give evidence of geological antiquity: Lake Posso in Celebes, Lake Lanao in the Philippines, and Lake Ochrida in the Balkans. The largest of inland bodies of water, the Caspian Sea, I have not mentioned since it occupies a peculiar place as an arm of the ocean cut off in rather recent geological times. It is a fascinating enough place from many different points of view, but its characteristics are too special to be relevant to our present discussion.

Lakes, in comparison with the sea, are shallow. The deepest fresh-water lakes are Baikal and Tanganyika, which reach depths of 5595 feet and 4707 feet. Lake Superior, the largest body of fresh water, has a maximum depth of 1008 feet and an average depth of 475 feet. The water of the clearest of lakes is not as clear as that of the open ocean: light penetration in one of the clearest, Crystal Lake, Wisconsin, is approximately the same as that in Puget Sound.

Once they have formed, lakes always tend to be growing shallower and smaller from the sediment brought into them by streams, or from the vegetation growing on their margins; or they tend to be drained away through erosion at their outlets. The shrinking process can be reconstructed easily enough in any particular area by comparing a series of lakes, ponds, swamps, marshes and bogs; one can see that many a fertile farm is located in the bed of some extinct lake. For the animals with amphibious lives this does not matter much; as one lake dwindles, they can move to another with more suitable conditions. For animals chained to the water, this is more serious. Their only escape is through the streams and the rivers. The aquatic fauna of geologically transient lakes thus tends to be different only in proportion from the fauna of the streams of the same region. It is necessarily part of the same system and there is little chance for peculiar forms to develop in a particular lake.

Rivers, lakes, ponds, marshes, with all of their varieties, still do not hold the entire story of life in fresh water. There is the curious special fauna of underground water and of streams and pools in caves. And there is another curious and special fauna that lives in water accumulated and held by plants.

The commonest plant containers of water are rot holes in trees. These are the only kind in northern forests, and there are a number of insects and other things that live only in this peculiar habitat. In northern bogs there is a plant, the pitcher plant, that accumulates water in its pitcher-like leaves. The plant is carnivorous, getting food from the decaying bodies of insects that have been trapped in its little horde of stinking water. But one species of mosquito successfully breeds in this water and nowhere else.

In tropical forests, these plant containers abound, and they have come to be inhabited by a quite special fauna. With almost daily rains, and with a high humidity that cuts down evaporation, water is caught and held in all sorts of places: not only in rot holes, but in leaves, flowers and fruit husks on the ground; at the bases of the leaves of several kinds of plants; and in certain types of flowers. The big spathes that shield the flowers of palms make fine water containers when they fall to the ground, and mosquito larvae, water beetles and the like teem in this water. Coconut husks collect water, as do the pods of cacao beans, and the hard shells of other large fruits.

The most abundant plant containers of water in tropical America are the bromeliads, the plants of the pineapple family. These are mostly epiphytic—that is, they grow on trees, not as parasites, but using the tree as a perch. There are many hundreds of species of bromeliads, and with most of them the bases of the leaves are broad and closely spaced, making a tank which holds water. A big bromeliad may have several quarts of water, inhabited by a wide variety of aquatic creatures. There is a frog that breeds in bromeliads; a dragon fly, many species of mosquitoes and other flies, beetles, worms and a host of microscopic organisms.

I became fascinated by these container habitats during my years of working in the Colombian rain forests. There were well over a hundred different species of mosquitoes in those forests, and finding out where each of these bred became a complicated game. We knew that most of them bred in plant containers. After we had exhausted the obvious places, we would go out into the forest and try to imagine where water might accumulate and then look for the water—and if we found water, we always found some species of mosquito using it!

The most curious of the container habitats, I think, was the water that accumulated in the internodes of growing bamboo. The bamboo would frequently have tiny holes made by some boring insect through which water would trickle and accumulate in the hollow stem. And in this water there would always be mosquito larvae—extremely queer larvae, so queer that they reminded me of the odd forms of deep sea fish. One kind of larva had a breathing tube much longer than its body, so that it looked like a tiny dog attached to an immense tail. Why this very long breathing tube, I don't know; other kinds of larvae in the same place would have very short tubes and seem to get along all right. One of these short-tubed fellows was very hairy, looking like a tiny woolly-bear caterpillar, which again seemed out of place in a mosquito larva.

But the oddest things about these mosquitoes were their habits. The entrance hole was often so tiny that I could not imagine how a mosquito could get in or out, yet the larvae would be there. One of these species in Panama, Pedro Galindo has discovered, hovers outside of the worm hole and shoots its eggs in with great force and accuracy. A species in Ceylon lays its eggs on its leg, and then sticks the leg into the hole in the bamboo.

It turned out that these bamboo-breeding mosquitoes were expert jail-breakers: ordinary wire netting would not hold them in at all. A normal mosquito flies up to a screen, bumps against it, and gives up. A bamboo-breeder, encountering a wire screen, sticks its head into the meshes, gives a few twists to its body, and wriggles through. After all, they have to be good at wriggling through small holes to get out of their breeding places.

But these mosquitoes, and the plants that hold the water in which their larvae breed, are a part of the story of the forest. Let's look next at some of the general characteristics of tropical forests.

7. The Rain Forest

Here Nature is unapproachable with her green, airy canopy, a sun-impregnated cloud—cloud above cloud—and though the highest may be unreached by the eye, the beams yet filter through, illuming the wide spaces beneath—chamber succeeded by chamber, each with its own special lights and shadows.

—W. H. HUDSON, in *Green Mansions*

My two favorite kinds of places in this world are coral reefs and rain forests. I don't know how I would vote if I had to choose between them, had to decide that I could go only to reefs and never to forests again, or vice versa. My idea has long been to live by a broad, sandy beach with a rain forest behind me and a coral reef off-shore before me, with either open to exploration or contemplation. Maybe someday I'll achieve the ideal. There are places where it is possible—some of the islands of the South Seas, or Trinidad and Tobago in the West Indies, for instance.

Certainly they are different enough, the rain forest and the coral reef. They have no inhabitants in common, nor even any general kinds of inhabitants in common. There is no way of comparing their appearance, either. The reef world is bright with color and movement. The forest is all green and brown, dim and still. The reef is Baroque, the forest, Gothic.

They have this in common: one is the product of the most favorable possible conditions for life in the sea, the other for life on land. Sunlight, warmth, moisture, are always abundantly present, stable, and favorable throughout the year. Moreover, they have remained about the same over long stretches of geological time. As a result, there is a tremendous variety of different kinds of organisms in both environments—and these organisms, among themselves, have developed a tremendous variety of different kinds of relationships.

"Rain forest" and "jungle" are frequently taken to mean the same thing. But I have never liked the word jungle. It has all the wrong connotations. You hack your way painfully through the lush vegetation of the jungle, dripping sweat in the steam-bath atmosphere; snakes hang from trees and lurk under foot; leopards crouch on almost every branch and there is always a tiger just beyond the impenetrable screen of foliage. There are hordes of biting, stinging and burning things. The jungle is green hell.

I doubt that there is any place, outside of books and movies, where all these conditions are combined, though certainly there is plenty of nasty and difficult country, both in the tropics and out. The thickest tangle of vegetation is the second growth that springs up after rain forest has been cleared. Everywhere in the tropics, people follow a slash-and-burn type of agriculture. Trees are felled, allowed to dry, burned, and crops planted among the charred logs. Sometimes crops are harvested for two or three years or more, but presently the land is abandoned and a new area cut. The abandoned clearing is taken over by a thick tangle of vegetation that for several years may be almost impossible to penetrate except by slow and painful cutting with a bush knife. Small mammals and rodents multiply in this vegetation, providing abundant food for the snakes that move in. Such places sometimes harbor swarms of mites, ticks, flies, mosquitoes and a wide variety of stinging things. It's about as close to a green hell as you can get.

The true rain forest, untouched, almost untrodden by man, is a very different sort of place. The forest floor is open, carpeted with the richly variegated browns of many different kinds of fallen leaves, sometimes brightly spotted with blue or red or yellow from flowers that have fallen from unseen heights above. The carpeting is thin, easily scuffed away to show the red lateritic clay soil so characteristic of the equatorial regions. There is no thick accumulation of leaf mold like that of northern forests, no rich accumulation of humus. The processes of decay are too fast to permit much organic accumulation in the soil.

There is little vegetation on the forest floor since the light is too dim for plants. There is a thin growth of tree seedlings (which have no chance to grow unless some catastrophe to a forest giant should open space), ferns, sometimes dwarf palms, or scattered thickets of huge-leaved aroids, the sort of plants that also grow well in the dim light of hotel lobbies. But basically, the forest floor is open, with vistas of a hundred feet or more, vistas framed and closed by the straight trunks of the trees that disappear into the vaulted green canopy that they support above.

The cliché often used for the forest is "cathedral-like." The comparison is inevitable: the cool, dim light, the utter stillness, the massive grandeur of the trunks of forest giants, often supported by great buttresses and interspersed with the straight, clean columns of palms and smaller trees; the gothic detail of the thick, richly carved, woody lianas plastered against the trunks or looping down from the canopy above. Awe and wonder come easily in the forest, sometimes exultation—sometimes, for a man alone there, fear. Man is out of scale: the forest is too vast, too impersonal, too variegated, too deeply shadowed. Here man needs his fellow man for reassurance. Alone, he has lost all significance.

The rain forest is perhaps more truly a silent world than the sea. The wind scarcely penetrates; it is not only silent, it is still. All sound then gains a curiously enhanced mystery. A sudden crack—what could have made it? An inexplicable gurgle. A single clear peal—that was a bird, probably a trogon. A whistle, impossible to identify. But mostly silence. The silence sometimes becomes infectious; I remember sometimes trying to blend into this world by moving along a trail without rustling a leaf with my feet or popping a twig. But more often I purposely scuffled, broke noisily through this forest where I didn't belong, tried to advertise my presence both to reassure myself and to warn the creatures of the forest that a stranger was there—I had no desire to surprise a fer-de-lance.

In contrast with the reef, it is a monotonously-colored world. Everything is some shade of brown or gray or green. I have lugged cameras loaded with color film all day without finding anything that seemed to warrant color photography, and in desperation, got my companions to wear red kerchiefs and blue jeans so that they could provide color contrast as well as "human interest." But I always found color photography difficult in the forest where the dim light requires long exposures, and the light itself is greatly altered by being filtered through the thick, green canopy. Only by taking advantage of the margins of clearings, or by using a flash, can you be sure of results.

Perhaps I am making the forest sound too easy, too open, too cathedral-like, overdoing my rebellion against the idea of jungle. It is difficult to give an objective description, to convey an accurate impression of a landscape like the rain forest which may, in one person, arouse awe and wonder, and in another, fear and hatred. P. W. Richards, in his book, *The Tropical Rain Forest,* has justly remarked that "tropical vegetation has a fatal tendency to produce rhetorical exuberance in those who describe it." The exuberance mostly tends toward the green hell side, but perhaps I have overdone the cathedral analogy.

I doubt whether the rain forest is anywhere easy to penetrate for any great distance. There are always obstructions: occasional fallen trunks, sudden tangled thickets, and above all, stretches of swamps and countless streams. Sometimes the streams are small, clear, shallow sandy brooks, looking no different from the forest brooks of New England, and easily negotiated. But sometimes they are broad rivers,.sometimes they move sluggishly over bottomless mud, sometimes they are choked with impenetrable masses of fantastic vegetation. The green hell analogy becomes vivid enough in these forest swamps. They are the reason that man has had so little success in making trails or roads through the forests; why he clings to the major rivers either for exploration or trade.

The world I have been describing bears no obvious resemblance to the world of the coral reef. This is partly, of course, because we have been looking at the forest from the bottom, from the point of view of walking on the forest floor, while we see the reef from above, floating over it. I loved the rain forest from my first encounter with it when I had just left college and was working at a research station of the United Fruit Company in Honduras, but I think I never really appreciated the forest until the years of yellow-fever study in Colombia, when I worked comfortably from platforms high in the trees—which gave a quite different perspective. But in any event, the similarities only become apparent when we examine the two habitats as biological systems, and for this we need a biological description of the forest.

Rain forest is the type of vegetation that occupies the lowland tropics in regions of high rainfall, where the rain is fairly evenly distributed throughout the year. The minimum rainfall to support such a forest is generally considered to be about eighty inches a year, though usually the rainfall is much higher, well over a hundred inches. The seasonal distribution is as important as the total amount: where there is a pronounced dry season of several months with little or no rain, the forest changes considerably in character and is generally called a monsoon forest. In such a forest the trees are not so tall, there is more undergrowth, and many plants drop their leaves during the dry season. The forest also changes with mountain altitude and various kinds of montane forests can be recognized. Heavy forests occur in some high rainfall areas outside of the tropics, in southern Chile, in New Zealand, and along the coast of the state of Washington in North America; but these again have a different character from the tropical forest and are best considered as a separate vegetation type.

There are three major areas of tropical rain forest: the American, the African and the Indo-Malayan. They cover all of the land masses crossed by the equator except the east coast of Africa. The American forest is by far the largest and most continuous, covering most of the Amazon drainage in central South America and extending south on the inner side of the Andes in Bolivia into the drainage of the Plata, and north in Colombia into the drainage of the Orinoco. This is hardly separated by the northernmost ranges of the Andes from a Pacific strip that follows the coast from Ecuador to Panama and continues, in Central America, along the Caribbean coast almost to the line of the Tropic of Cancer in Mexico. There is an isolated stretch of rain forest along the southern coast of Brazil, and rain forest once covered much of the West Indies, though now there are only scattered remnants.

The African rain forest is the smallest of the three, and there is considerable debate both about its present limits and its former extension. It covers, essentially, the central drainage of the Congo, with a north and west extension along the gulf of Guinea to Liberia.

The Indo-Malayan rain forest is the most fragmented. It covers most of the large islands of the East Indies—Sumatra, Borneo, Celebes, New Guinea, the Philippines—and the Malay Peninsula, with outlying areas on the west coast of India, in Burma, on the coast of Indochina, and along the coast of northern Queensland in Australia.

In structure and appearance, the forests of the three areas are very much alike. The taller trees reach a height averaging about 150 feet, though individual trees more than 200 feet in height are not uncommon. The tallest reported rain forest trees are somewhat less than 300 feet. Rain forest trees are thus in general taller than trees in the temperate forests of Europe or North America, where the average for taller trees in the least disturbed forests is around 100 feet, with 150 feet an exceptional height. But trees in the tropical forest do not reach the gigantic proportions of the California redwoods or the Australian eucalyptuses.

The tallest measured sequoias reach 364 feet, the tallest eucalyptuses, 350 feet.

The rain forest everywhere has a multi-storied canopy. The layers of the canopy, like the depth zones of the sea, are hard to define since they are not sharply separated, but it is customary to refer to the upper, the middle and the lower zones of the canopy. This multiplicity of tree layers reflects the great variety of different kinds of trees that go to make up the rain forest. Most temperate zone forests are made up of one or two or a very few kinds of trees, with a maximum, in the Appalachian forests of the eastern United States, of perhaps 25 species of trees. Probably fifty species of trees is about the minimum for any rain forest, and in most places there are many hundreds of species.

Alfred Russel Wallace, who spent many years exploring both the Amazonian and Malayan regions wrote, in his book *Tropical Nature:* "If the traveller notices a particular species and wishes to find more like it, he may often turn his eyes in vain in every direction. Trees of varied forms, dimensions and colors are around him, but he rarely sees any of them repeated. Time after time he goes towards a tree which looks like the one he seeks, but a closer examination proves it to be distinct. He may at length, perhaps, meet with a second specimen a half a mile off, or he may fail altogether, till on another occasion he stumbles on one by accident."

Because of this immense variety, the catalog of rain forest trees is still far from complete, and anyone collecting botanical material in remote areas is liable to turn up species of trees unknown to science. And the trees belong to almost the whole range of plant families. Families that in the north are known only as herbs—the *Compositae,* or daisy family, for instance—are represented by trees in the rain forest. Even grass takes on a tree form in bamboo; and ferns, in the tree ferns.

There is a great development of woody vines—lianas. A considerable proportion of the foliage in the canopy, sometimes nearly half of it, is from the great vines that are supported by the trees, and these woody vines, like the trees themselves, belong to a great variety of species and to many different plant families.

Trees, woody vines and epiphytes are the characteristic life forms of plants in the tropical forest; and of these the epiphytes, the plants that perch on the branches and trunks of trees, are probably the strangest to the observer from the north. Lichens and mosses grow as epiphytes in northern forests, but in the rain forest the branches and trunks of the trees are covered with a bewildering variety of other plants: ferns, orchids, peppers, cactuses, bromeliads. In all, something like 33 families of seed plants and ferns are represented by epiphytes in the rain forest flora.

Epiphytes grow in the rain forest, but without access to the soil they are faced with a water problem. Their niche on dry branches high in trees is in a way a sort of microdesert. Many of them, the cactuses, for instance, are relatives of typically desert plants and have the fleshy look of desert plants. There is a good reason for this, since in both situations it is important for the plant to be able to conserve water, hence the frequency of succulent bulbs and leaves, as in the orchids. The bromeliads, as I mentioned in the last chapter, have solved the water problem by forming watertight tanks where they can collect their own supply, and where also they can get food from the rotting organic matter. The bromeliads are often very numerous, forming, as someone has remarked, a "marsh in the treetops."

The epiphytes are also faced with mineral problems. They must get the salts they need from the extremely dilute solutions in the rain that washes them, or from the humus and debris that collects in the cracks of the bark of the host trees or in the tangle of their own roots. Somehow they get sufficient minerals from these sources. Frequently the epiphytes live in a close symbiotic association with fungi, and the thin mycelial webs of the fungi help both partners in food collection. Fungi live in close relationships with the roots of many kinds of seed plants, forming associations that are called *mycorhizae.*

The roots of epiphytes also, with surprising frequency, are used as nesting sites by the ants that abound in the forest; and it has been suggested that the epiphytes have an additional source of food in the material that the ants are constantly hauling into their nests. This too would be a symbiotic relationship: the ants get a fine, well-protected home, built by the plant, and provide food by way of rent. Perhaps they also provide defense since many of these ants sting fiercely. They certainly defend the epiphytes valiantly against stray humans trying to make collections.

The rain forest crawls with ants of many different kinds, occupied with many sorts of business. Close associations between particular kinds of plants and particular species of ants are quite frequent. Sometimes the ants clearly serve to protect the plants that provide them with nesting sites in hollow stems. One learns to avoid brushing against the trunks or foliage of certain trees, like Cecropia, with the same care that one learns to avoid poison ivy or poison oak in more northern situations. With the ant-protected trees the fiery consequences of transgression are immediate as well as painful.

The ants are one example of the incredible abundance of insects in the rain forest. It is an abundance of kinds, rather than of individuals. Within a range of about ten miles of our laboratory in Colombia, we found 150 different species of mosquitoes (there are only 121 species known from all of the United States and Canada). But you may get more mosquito bites in northern woods than in tropical forests. In northern woods, the mosquitoes biting you are apt to be all the same kind; while in the rain forest, almost every bite will be from a different species of mosquito—if that is any comfort.

The task of collecting, naming and describing all of these insects is endless. Every collector brings back new things and no one really has any idea how many different kinds there are, which leads people, in guessing about how many species of insects there are in the world, to give figures that vary from one million to ten million. Whatever the figure, it is a big one, with a respectable proportion inhabiting the rain forest.

The biggest known insects are found there: for wingspread, butterflies and moths; for bulk, rhinocerous beetles; for length, walking sticks. But there are relative giants among almost all insect groups: cockroaches that look like small turtles, big flies, big wasps, monstrous grasshoppers. There are also big mammals, elephants in the old world, tapirs in the new, and there is a gigantic frog in the Congo forest with a body approximately a foot long. But the woolly mammoths that roamed Europe, Siberia and North America until almost recent times were bigger than elephants, and in general with mammals, and birds, the forest representatives do not strike one as particularly big. As a matter of fact, forest species—forest deer, for instance—tend to be somewhat smaller than their cousins in the savanna or outside the tropics. Perhaps it is only the cold-blooded animals—anacondas and boas and pythons, for example—that find a special opportunity for bigness in the rain-forest environment.

The color is mostly high in the canopy: the flowers of the trees, lianas, epiphytes; birds and butterflies. There is, it seems to me, a special tendency for animals, especially birds and butterflies, to be colored in metallic blues and greens. The wings of a great morpho butterfly, flashing in the sun, are especially famous, but many kinds of day-flying insects have metallic colors, even mosquitoes! Species that fly at night and species of the forest floor zone are, however, apt to be dull or at least softly colored. With the butterflies, as with the birds, the brilliance tends to be a male characteristic.

The structure of the rain forest and the appearance of its inhabitants is much the same in America, Africa and Indonesia, but the three areas have had independent evolutionary histories, and are made up of quite different kinds of plants and animals. The only animal I can think of (besides man) common to all three regions is the leopard, which is hardly distinguishable from the jaguar of the New World. But though these big cats are at home in the rain forest, they are far from restricted to it, or to the tropics either for that matter. The leopard, in prehistoric times, ranged far into Europe, and the jaguar ranged all over both American continents until man managed to exterminate it in the settled regions.

The African and Indo-Malayan forests are more similar to each other in composition than either is to the American forest. They are separated now by the Indian Ocean within the tropics, and by the barriers of the Sahara and the Himalayas to the north. But the Himalayas are the youngest of mountains and the Sahara was once wooded, so that a few million years ago communication between Africa and Malaya was easier than now. But the Amazonian forest, all this while, was completely isolated by a sea barrier, only recently bridged by the isthmus of Panama.

Thus we find great apes in both the African and Indo-Malayan forests, gorillas and chimpanzees in the former and orangs and gibbons in the latter, but no great apes in tropical America. There are monkeys in all three regions, but the American monkeys belong to quite different families from the Old World monkeys, and have had a separate evolutionary history for a very long time.

Visitors to the tropics who expect to see monkeys everywhere are generally disappointed. In the rain forest, particularly, what one will see is very uncertain. The animals are there, all right, but they are not exhibitionists. They have learned, particularly, to be shy of possibly harmful human intruders. That this is learning, in part at least, is shown by animal behavior in places where the fauna has been long protected—in the park reservations and on the lands of Buddhist monasteries in the East.

One of the very few fully protected rain forest areas in the American tropics is a small island in Gatun Lake in Panama called Barro Colorado which was set aside in 1922 as a reserve to be used only for biological study. I only lived there for six months, and I probably saw more forest mammals in that period than in all my years of residence in other parts of the tropics. Mostly you get only a fleeting glimpse of a band of monkeys as they take off rapidly through the treetops—they have spotted you first, and are having none of it. On Barro Colorado, the monkeys did not really become pals with the visiting scientists, but at least they didn't panic at the sight of a man. As a result, any time you went out on a trail, you were sure presently to come across a band of monkeys and could watch them for a while.

I think, however, that I got more pleasure out of the coatis than out of the monkeys during this period on Barro Colorado. Coatis are long-nosed tropical relatives of the raccoons—and they are even more charming, inquisitive and intelligent than their northern relatives. They are hated by hunters because, ganging up, they can kill the hunters' dogs, and they are not notably careful about poultry or gardens. Men and coatis are thus usually at odds, but on Barro Colorado they achieved a considerable mutual tolerance. If you stood quietly, the coatis would pay little attention to you while they went on about their business of rummaging among the leaf litter or scampering up the trees after insects and nuts.

Frank Chapman was there while I was, carrying on his life-long study of bird habits. He tried to devise a feeding station for birds that the coatis couldn't get at by hanging a cigar box from a wire trolley strung between a tree and the verandah of his little house, so that bananas, for birds only, could be suspended in midair. It didn't take a neighboring coati long to figure that out. The coati, sniffing the banana from the ground, rapidly made for the tree, climbed it, did a tightrope act on the wire out to the cigar box, and got the banana. Chapman gave up and spent the rest of the time trying to fool the coati, with little success. He has told the story in his book, *My Tropical Air Castle*. As far as I know, coatis have not been investigated by psychologists, which seems a pity because they must have one of the highest I.Q.s in the animal kingdom.

Among the rain-forest animals, there are many survivors from the geological past. We tend to think of marsupials as primitive mammals that have survived and proliferated in the isolation of the Australian continent, except for the common and tough opossum of North America. But in the tropical American rain forest, there are dozens of kinds of marsupials. They are not as spectacular as the kangaroos and their relatives of Australia, but they are interesting enough. There is a sleek water opossum (*Chironectes*) living on the margins of the forest streams; a bright-eyed woolly opossum (*Caluromys*) with lovely, thick fur; and many different species of tiny, mouse-like opossums (*Marmosa*) as well as several other genera. The sloths are another archaic group (and they look it) now confined to the trees of the American rain forest, though in the recent geological past there were also many kinds of ground sloths.

The catalog of ancient animal types surviving only in the rain forest is a long one, though most of the items on the list can be appreciated only by the zoological specialist. Because of this, the rain forest has sometimes been regarded as a sort of backwash, a refuge from the more strenuous and progressive parts of the biosphere. But the whole complex and wonderful system of adaptations within the forest is clearly a consequence of evolutionary forces operating within this environment. Thus, by another chain of reasoning one can come to the conclusion not that the rain forest is a backwash, but that it is the place where evolutionary change is most active.

It is in the rain forest that "jungle law" reigns supreme: the struggle for existence, nature red in tooth and claw. Here we find the most fantastic contrivances for catching food or for avoiding becoming food, the most perfect examples of animal camouflage. The struggle for existence is symbolized by the strangling fig which starts out as an epiphyte, a seedling growing high on some tree, sending down roots which reach the ground and grow until finally the host tree is smothered by the encircling fig, which then stands alone. And I can think of nothing more devastatingly fierce, more irresistible, than a horde of army ants on the move, killing and dismembering every animal they encounter that cannot fly away or run fast enough. We had a snake pit in the laboratory where I worked in Honduras, where we kept many kinds of vipers for their venom, which was used for making antivenin. Once, I remember, we were invaded by army ants. We tried everything we could think of to stop them or at least to make them change their course, but to no avail. The ants poured on in their tens of thousands, swept through our snake pit, and left us with a collection of bare skeletons.

Life, then, can be grim enough for the forest inhabitants. Only the clever, the carefully protected, the extremely prolific, the most modern, seem to have any chance of survival. Yet, scuttling through the leaves of the forest floor, and nestling in the debris collected around the roots of epiphytes high in the trees, are scores of kinds of cockroaches hardly different in any way that we can see from the fossils of their ancestors that lived three hundred million years ago in the forests of the Carboniferous period. Push over a rotting log and you may very well find a Peripatus: a soft, brown, delicate, multilegged, caterpillar-like thing that, on examination, turns out to be a very queer creature indeed. The ancestors of all the land arthropods, of the millipedes, centipedes, spiders, insects, the first animals to learn to live on land, must have been something like this. But this anachronism is still getting on very well in the warm, damp world of the forest, meeting (as far as we can see) unchanged the shifting hazards in hundreds of millions of years of forest life.

Of course, neither the rain forest nor the coral reef has any monopoly on anachronisms. The horseshoe crabs of the Atlantic coast of the United States might have crawled out of the most ancient seas; the tiny collembola that jump through the leaf mold of northern woods are hardly different from the earliest of insect fossils. Every pond and every tidepool teems with survivors from the ancient past. I don't think we have any clear answer as to why some kinds of things have survived and other kinds perished as, with the onward flow of time, new kinds of organisms have come on the scene. Our explanations are mostly circular: horseshoe crabs have survived because they were adapted or adaptable, that is, able to survive; trilobites became extinct because they weren't able to survive. One of the remarkable things is that very few of the major plant or animal types —phyla or classes—have become extinct. Proportions change —ferns and mosses give way to seed plants as trees, giving structure to the forests—but the ferns and mosses are still with us. Reptiles give way to mammals and birds as dominant animals in the forest scene, but reptiles are still with us. Change seems to be not so much a matter of absolute replacement as of diversification.

But with all these reservations and qualifications, the rain forest still seems to have more than its fair share of survivors from the past. One can visualize the struggle for survival, the competition, the strenuousness of life in the rain forest. But then in the next instant, looking at this multitudinous accumulation of organisms one gets the feeling that there is so much warmth, so much light, so much moisture, so much food, that almost anything can survive and that almost everything does.

This is the paradox shared by the rain forest and the coral reef. Shallow, tropical seas and low, rain-drenched tropical lands with their associated swamps, lakes and rivers represent the most favorable conditions for the processes of life on the surface of our planet. The physical environment is not a problem except in special situations like that of the epiphytes of the forest where the collection of water or minerals may have particular difficulty. The problems of living things involve not so much the physical environment as other living things. When one moves away from the optimum of these two environments, the physical condition starts to become limiting. The open sea has a less diversified biota than the reef because it has no solid substrate, no place where fixed organisms can grow; everything has to drift or swim. Descending from the surface of the sea, light and temperature become less favorable; going toward the poles in surface waters, temperature becomes less favorable. Similarly, on land, as one moves from the rain forest to other types of habitat, various factors of the physical environment become limiting, present special problems which require special adaptations on the part of the organisms living in the habitat. Organisms must always cope with other organisms, but outside of the rain forest and the coral reef they must increasingly cope with climate too. We'll look at some of the adaptations required by these different land climates in the next chapter.

8. Woodland, Savanna and Desert

Every farm woodland, in addition to yielding lumber, fuel and posts, should provide its owner a liberal education. This crop of wisdom never fails, but it is not always harvested.

—ALDO LEOPOLD, in *A Sand County Almanac*

The rain forest, the coral reef and the open sea are remote places. Our encounters with the living world are in our gardens, in our parks, in the roadside woods where we stop for picnics, in the forests and streams where we hunt and fish. Western man—any civilized man—has a curiously limited contact with the natural world.

The remoteness, in the case of the rain forest and the coral reef, is geographical. In the case of the open sea, the remoteness is intellectual. We may travel across the sea often enough, but we are hardly experiencing the environment. We lounge in a deck chair, staring at the horizon. Or we lean over the rail to watch the play of waves; sometimes to see flying fish skittering out from the disturbance of the bow; or to wonder at a school of porpoises playfully pacing us. But we are living on a bit of land that has been built to float us from one continent to another. The canoes of the Vikings and the Polynesians, the caravels of Columbus, the clippers of the New Englanders, hardly brought their occupants any closer to the sea. These bits of land were smaller, less stable, more vulnerable to wind and waves—but the human was still remote from the unseen plankton, from the armies of squid, from the unimaginable world of bizarre fish and flashing lights in the depths. We can only know this world by an act of imagination, reconstructing it from many bits of evidence given us by microscopes and towing nets and bathyspheres.

I think the world of inland waters is similarly remote in many ways, even though here we are in much more frequent and more direct contact. When we fish a trout stream we may become very expert in choosing flies and in spotting likely pools, but do we really have any understanding of the world in which our trout is living? And that pond of mine again: I love to watch the dartings of the minnows, the huddling tadpoles in the puddles of the swampy edge, the bright-eyed turtles that stick their heads out to check on me at a safe distance. But always I am an outsider watching a play, trying to find its meanings. The surface of the water, like the footlights of the stage, is a barrier I cannot cross —except to kill, pollute, and disrupt.

The world of western man is the world of the woods and grasslands of midlatitudes. This, I suspect, is not too far from the original human habitat. It seems likely that man became man—acquired his physical character and the beginnings of his culture—at the margins of the tropics, in places where the rain forest gave way to less overwhelming landscapes. And man has continued to flourish there, though penetrating an ever wider variety of environments, as he develops the cultural equipment to cope with conditions that otherwise would be adverse.

Perhaps, then, we should start any examination of the biosphere with this familiar world and work gradually toward the strange creatures and events of the equatorial forest or the ocean depths. There are two difficulties in this, I think: first, that we can hardly see what lies under our noses; and second, that man has so drastically altered the landscapes he can control that it is hard to disentangle man from nature.

The first difficulty is real enough. We tend to take the familiar for granted and to wonder at the strange and new. To see the familiar, we have to have some sort of a new way of looking at it, some sort of a revelation, perhaps, or new insight or perspective.

My first synaptid sea cucumber started me thinking about this. It was on Ifaluk, that atoll in Micronesia where I spent a summer prowling, or rather floating, about the reefs. There it was, a long, slender, mottled, snakelike animal extending out from under a coral boulder, looking like a misplaced fer-de-lance. But when you looked for the wicked head, you saw a small, gently waving bouquet. And when you reached down to pick it up, it collapsed—the thing was made of tissue paper.

"What is it?" I asked Don Abbott, who was with me.

"Oh, that's a synaptid sea cucumber. It's one of the improbable animals."

"What do you mean by an improbable animal?"

"It's the sort of thing I never would have thought of if I had been giving advice about animal creation," he said.

We kept finding improbable animals all summer. It was easy on Ifaluk. Then, back in my study in Michigan, I started to write an article about improbable animals: all the strange creatures that didn't act or look the way one thought they should. But then I thought about all the things in a drop of pond water—they were improbable, too, because you ordinarily couldn't see them. But a squirrel or a rabbit becomes improbable if you look at its hair or a piece of its flesh with a microscope—if you see it with a new perspective. The more I thought about the different animals in my own back yard, the more improbable they all seemed. In the end, I decided that everything was improbable and gave up the idea of the article—but I had to go to Micronesia to get this particular view of my back yard.

The second difficulty in dealing with the landscapes of Europe and North America—in dealing with landscapes anywhere that man has settled thickly—is in separating human effects from natural effects. This sounds as though man were unnatural, which turns on a matter of definition that I would like to avoid for the moment. Natural or not, the biotic community wherever man has settled thickly is clearly different from the community that would exist in the same place without man, and for many purposes it would be useful to know about the distribution and interrelations of organisms in the absence of human influence. This leads ecologists, in charting and describing the landscapes of the world, to play the game of "let's pretend that man doesn't exist." It is a complicated and somewhat difficult game that at times leads to considerable bickering among the players, but it is an important game for the understanding of our world.

There are many kinds of clues useful in playing this game. Most important are the patches here and there that apparently have escaped direct or indirect modification at the hands of man, like the remaining redwood forests of California and other western forests in North America and, less clearly, fragments of the Appalachian forest. In the case of North America, and other parts of the world where man was inconspicuous until modern times, we have the accounts of the first explorers—sometimes tantalizingly inadequate, sometimes exaggerated, sometimes extremely useful. John Bakeless has reviewed what we have been able to learn about pre-European North America from such sources in *The Eyes of Discovery,* which shows the possibilities of this approach.

Another important kind of clue is through plant pollen. Each species of plant has characteristic pollen, which can be identified by microscopic examination. Most forest trees are wind-pollinated, producing great quantities of pollen annually which settles everywhere and which in some circumstances may be preserved indefinitely. Preservation is particularly good in peat bogs, and by examining the proportions of different kinds of pollen preserved at different depths in a bog, the composition of the vegetation of the surrounding areas in past times can be reconstructed. We can get, by this means, a history of the forests of various parts of Western Europe and North America over many thousands of years. Primarily this gives us a record of the changes in vegetation resulting from the changes in climate as the glaciers retreated, but in Europe it also gives us a record of the first human influences.

From such studies, it is clear that the eastern United States, the British Isles and Central Europe, and much of China were formerly covered by vast forests, similar enough in general characteristics to be grouped as a single type, which the ecologists call the "temperate deciduous forest." In Europe and China this forest was greatly modified by human activity long ago, but in North America the scattered clearings of the Indians had little effect so that the landscape encountered by the first Europeans was the "natural" landscape, hardly touched by man.

This forest, in midsummer, was similar in many ways to the untouched tropical rain forest. Thomas Ashe, who traveled through Pennsylvania in 1806 noted that "the American forests have generally one very interesting quality, that of being entirely free from under or brushwood. This is owing to the extraordinary height, and spreading tops, of the trees; which thus prevent the sun from penetrating to the ground, and nourishing inferior articles of vegetation. In consequence of the above circumstance, one can walk in them with much pleasure, and see an enemy from a considerable distance."

The deciduous forest was also, it seems, a silent world by day. The various songbirds so common in our modern woods were then confined to the forest margins, the coast, river valleys, the occasional areas of grass and scrub. The birds of the forest were owls, ravens, eagles, pigeons, the Carolina parroquet—species that in many cases have become rare or extinct with the disappearance of their habitat. Many accounts agree that there was plenty of noise in the forest at night: hooting owls, howling wolves, screaming pumas. They were mostly unearthly, frightening noises. Pumas, common enough in those days, produce a sound which reminds me of what might result from slowly murdering an extremely vocal baby.

This deciduous forest had nothing like the variety of trees of the rain forest, but the trees were not uniform stands, either. Usually several species were common, occurring in particular combinations like beech-maple, oak-hickory, or elm-ash-maple. The trees were tall, but not as tall as in the tropical forests, and there was nothing like the variety of vines and epiphytes found in the tropics. And the nature of the forest was governed, of course, by the fact of winter, by the annual period when all growth stopped, when the trees dropped their leaves.

The traces that are left of this once vast forest are mostly in remote mountain areas, in the Appalachians of the United States and in southern China. Man has been extremely successful everywhere in the territory of the deciduous forest and he has remade the landscape for his own purposes, turned it into fields and orchards or sometimes into forests carefully managed as a human resource. In the United States, many areas have been lumbered and then abandoned to allow a new tree growth to start unplanned by man; or fields, once cultivated, have been left to grow up again with trees. But the woods in these cases are quite different from the original forest. For one thing there has not been time for trees to reach the giant size of their predecessors. For another, the kinds of trees occur in quite different proportions from those found in the primeval forest. There is a sort of instability about the secondary growth of trees: some kinds, like white pine, paper birch, red maple, may be able to establish themselves most easily in the grass or weeds of a cut-over area, while others like oak and hickory can only start at a later stage.

The instability of the cut-over landscapes so common in Europe and North America has led ecologists to be much concerned with the "succession" that leads up to a final stable, or "climax" community. They distinguish between primary and secondary succession. The first occurs as a pond slowly fills up to become a bog, a meadow, and finally a forest; or as a sand bar formed by a river slowly becomes consolidated with the dry land of the banks. The second occurs when man, purposefully or accidentally, destroys a stable community. Sometimes a sort of artificial stability is achieved through human action, as with the pine woods of our southern states. These seem stable enough, but they are maintained only through periodic fires which kill the oaks and other broad-leaved trees that otherwise would eventually replace the pines; these pine woods, for the ecologists, are a "subclimax."

To the north of the deciduous forest in North America, Europe, and Asia lies a broad band of evergreen, coniferous forests. Ecologists use a Russian word for this landscape, calling it a *taiga*. Man has made considerable inroads into the taiga, but it still includes the largest areas of forested wilderness left on our planet. The trees are mostly spruce, fir and pine of various species, with thickets of alder, birch and juniper. When the taiga is burned or cleared and abandoned, aspens and birches come in first, but in the long run they are again replaced by conifers. The slow decay of the needle litter results in a characteristic soil type for the taiga region, again called by a Russian word, *podzol*. The moose is the most conspicuous animal of the taiga, and the forests have a characteristic set of smaller mammals: black bears, wolves, martens, lynxes and many kinds of rodents.

The transition from deciduous forest to taiga is largely controlled by temperature, by the increasing severity of the winters. Still farther north, as average temperatures become lower and the growing season becomes shorter, the taiga gives way to the treeless *tundra*—another Russian word. In the tundra, the ground is permanently frozen a few feet below the surface, and frosts are liable to occur at any time during the summer. The plants and animals that grow there thus require quite special adaptations, and the number of species of organisms is consequently greatly reduced, though individuals may be abundant enough—flies, mosquitoes, flowers, mosses, reindeer, birds—so that on warm sunny days the tundra teems with life. The taiga extends south in the mountains, represented by the coniferous forests of the Rockies and the Himalayas. And conditions above the tree line on mountains are similar to those of the tundra.

Changes in the living landscape from south to north, then,

in the northern hemisphere, are primarily controlled by the changes in average or minimum temperature. Moving from east to west at a given latitude—and elevation—we are apt to find the changes controlled by rainfall, and the sequence is woodland, grassland and desert. It isn't as simple as this, of course, but one can say that the type of vegetation is in general governed by moisture and temperature, and that the nature of the whole biological community—animals and plants—depends on the type of vegetation. In high latitudes, temperature is most important; in midlatitudes and in the tropics, the influence of temperature is always modified by moisture.

The seasonal distribution of rain is important, as well as the total annual amount. The deciduous forest depends on adequate rain during the summer. Around the Mediterranean and along the California coast there is a special climate with winter rains and summer drought which has resulted in a special type of vegetation. The trees retain their leaves throughout the winter, when most growth occurs, and are adapted to withstand the dry heat of the summer through deep root systems and foliage shaped to reduce evaporation. Holly, cypress, olive and similar trees made up the forest that originally covered the coastal plains and lower mountain slopes of the Mediterranean—and it was a real forest, though probably never as imposing or dense as the deciduous forest farther north. Man destroyed this forest long ago, but the climate and the resulting vegetation formation have a special interest because of the importance of the Mediterranean lands in the history of Western civilization.

There are many references to forests among classical writings. Homer spoke of "wooded Samothrace" and "wooded Zacynthus" and of the "tall pines and oaks of Sicily." Forest fires in the dry Mediterranean summer must have been spectacular. Thus Homer wrote that "through deep glens the fierce fire rages on some parched mountain-side, and the deep forest burns, and the driving wind whirls the flame every way." Thucydides wrote of "spontaneous conflagrations sometimes known to occur through the wind rubbing the branches of a mountain forest together." As the forests were cleared, the goats took over, and goats have been a dominant factor in determining the nature of the Mediterranean landscape ever since. The devastation was already clear at the time of the height of Athenian glory and Plato saw that "What now remains compared with what then existed is like the skeleton of a sick man, all the fat and soft earth having been wasted away, and only the bare framework of the land being left."

The parts of California with a Mediterranean-like climate are covered, when not cultivated, with a growth called *chaparral.* The present chaparral is greatly influenced by human activity. Many of the common plants are exotic, purposefully or accidentally introduced by man; others are new hybrid forms of the sort that develop where man removes the barriers that prevent the mixing of isolated, local species. And the present chaparral equilibrium, like the equilibrium of the southern pine woods, is the consequence of periodic fires. What the growth would be like in this region in the absence of human interference is not clear.

Mediterranean forest, scrub, grassland, desert, form a sequence of vegetation types that cannot be sharply differentiated. All are characteristic of regions where water is either scarce or seasonably unreliable, but the details of vegetation in such circumstances vary endlessly, depending not only on the rainfall, but on temperature, soil conditions, geological history and recent human activities.

The word "desert," for instance, covers a considerable variety of biological conditions from regions like the mid-Sahara where water is completely unavailable, to regions with a rainfall of as much as ten inches in a year—at least ecologists generally draw the line at ten inches of rain in defining desert conditions. But despite the difficulty of definition, we all know what a desert is: it is a place where organisms cannot live because of the absence of water, or where the organisms that do occur, like camels and cactuses, have special means for dealing with water problems.

There are seven great desert areas in the world: the Sahara in North Africa, the Kalihari in South Africa, the Arabian and

Thar in Asia, the Victoria in Australia, the Atacama-Peruvian in South America, and the Colorado-Sonoran in Mexico and the southwestern United States. These cluster around the northern and southern lines of the tropics—as the rain forests cluster around the equator. The deserts lie in the horse latitudes, in the regions of the trade winds which, blowing toward the equator, from cooler to warmer latitudes, are drying winds unless they happen to meet the cooling obstacle of the mountains. These deserts merge into grassland and scrub on both sides, toward the equator and toward the poles, except that in the southern hemisphere—in Australia, Africa and South America—the land mass is not large enough to permit as full a sequence of climates and vegetation types as in the northern hemisphere. Still, the pampas of Argentina and the veld of South Africa correspond to the prairies of North America and the steppes of Europe and Asia in their major biological characteristics, as well as in the ways in which they have been utilized by man.

These great deserts of the world are separated, north and south, by the wet tropics; east and west, by the oceans. This means that there has been relatively little chance for interchange of desert-adapted animals and plants among them. Each of the desert areas has acquired, more or less independently, a set of inhabitants able to cope with these severe conditions for life. This has sometimes resulted in remarkable similarities among quite unrelated organisms. The cactuses, for instance, are a purely American family of plants that have become adapted to arid conditions through a great thickening of the stems, which serve for water storage, and through various devices for cutting down water-loss through evaporation, including (in most cases) the loss of leaves, which are replaced by hard spines—the photosynthesis function being take over by the enlarged stems. In South Africa a group of quite unrelated plants, the euphorbias —relatives of the spurges and poinsettias—have achieved similar modifications and have assumed a completely cactus-like appearance in response to the desert conditions of that continent.

Desert plants in general tend to have large root systems. The mesquite bushes of the American deserts may have roots 30 to 100 feet long, penetrating to the underlying ground water. How plants like the mesquite get their start in life remains a mystery, though it is clear that once established, they may live for a very long time. Desert plants also tend to be tough and thorny or, like sagebrush, to have acrid juices. This, it has been suggested, is because devices for protection against plant-eating animals become particularly important under desert conditions, where plants are few and where they cannot readily become established.

These remarks apply to the perennial desert plants. There are, in addition, many kinds of plants that appear briefly in every desert after a heavy rain, growing rapidly, flowering, producing seed, and disappearing—sometimes not to reappear again until another appropriate rain, several years later. Fritz Went, of the California Institute of Technology, has studied the behavior of the seeds of such plants. They will only germinate under particular conditions—a rain of an inch or more in the months of November or December, in the case of Death Valley species. The seeds will not germinate if soaked in water; nor will they germinate if watered with less than an inch of rain. Dr. Went explains this in terms of "inhibitors," materials like salts in the soil that must be leached away by a certain quantity of water filtering *down* through the soil. Except for this highly specific germination requirement, and for the subsequent rapid growth, there is nothing very peculiar about this class of desert plants; they look very like their relatives in less rigorous environments.

The animal life of deserts is also restricted and specialized. Desert animals are small—gazelles, antelopes, coyotes and the like are characteristic of the semiarid scrub and grassland rather than of true desert, though this may be a matter of fine discrimination of what is "arid" and what is "semiarid." Of mammals, some rodents are able to withstand extremely dry conditions. The kangaroo rat of the southwestern United States is apparently able to "manufacture" its own metabolic water from the dry food it eats. Like many insects, of which the clothes moth is the classic example, it is able to live without any obvious source of water.

As one moves from desert conditions to scrub and grassland, the number of animals increases greatly. The animals of the prairies and savannas of the world are—or were—both numerous and spectacular: the vast herds of bison on the American prairies; the hordes of antelope, gazelles and zebras of the African veld; the guanacos of Patagonia; the wild horses of the Asiatic steppes. Such grasslands have possibly supported a greater bulk of animal life than any other kind of landscape. Most of the vegetation of forests stays as vegetation, decaying and returning to the soil without ever having gone through an animal cycle, but the grasses of the open plains of the world mostly become animal fodder, supporting both the herds of herbivores and a great variety of carnivores that live, in turn, on the grass eaters.

There is a great argument about the extent to which the grassy plains of the world are maintained by fires set by man, which limit the development of scrub and tree growth. The grasslands both of Africa and South America have certainly been greatly extended by this means within historic times, and grasslands everywhere give evidence of an uneasy equilibrium, of being a "subclimax" as the ecologists would say. But the grazing mammals of the world are evidence enough that grasslands have been in existence for a very long time to permit the evolution of all of the special characters of habit and physique needed for this existence.

The hordes of grazing animals themselves, like fire, serve as a force in maintaining the nature of the landscape. Such shrubs and trees as might get started among the grasses are nipped in the bud and the herds act as a sort of biological lawn mower which eliminates everything except the resilient, persistent grasses. Overgrazing, as cattlemen have learned, is possible: even the grasses eventually give up under constant eating and trampling, the soil becomes dust and the animals starve. In the natural system, this is prevented by the carnivores—the lions, tigers, and wolves of the world—which keep the numbers of herbivores within reasonable limits.

There is ample evidence of the importance of the carnivores for the well-being of the herbivores from the various instances where man has accidentally or intentionally removed the carnivores. An oft-cited case is that of the deer of the Kaibab Plateau of Arizona. A campaign was started in 1907 to exterminate the pumas, wolves and coyotes in the region with the idea of helping the deer by removing their enemies. The campaign was quite successful and the deer increased enormously in numbers, reaching a maximum in 1924; a maximum, it turned out, far beyond anything the range could permanently support. It is estimated that in the winters of 1925 and 1926 more than half the deer starved to death. The numbers continued to decline through starvation, though more slowly, over the next ten years, and the range continued to deteriorate. In cases like this, where man has removed the predators, he has to take over their function by killing the excess population of grazing animals—though man doesn't seem to achieve as nice a balance as the pumas and wolves.

This sounds fine when you read about it, but I am always bothered by the question of what keeps the predators under control when man isn't around. The classic case, and the case where this still can be studied, is in the grasslands of Africa. From the days of the first European penetration of that continent, all reports have stressed the immense numbers of game animals, of antelopes, gazelles, hartebeests, zebras, and the like, grazing in the open grasslands. The numbers have been greatly reduced through the activities of modern man, but the herds persist in many places, especially in the great park reservations. There the hordes of antelope live in healthy numbers amid the groups of sleek and well-fed lions. The lions keep the antelope from becoming numerous enough to ruin the grass, but what keeps the lions from becoming so numerous that they kill off their own antelope food supply? I keep asking the question of friends who might know, but the answers are not very clear. Yet it is the sort of question that might be answered by a keen observer carefully studying the relations among the animals in these great parks.

There is an interesting tendency among the animals of the open grasslands to form herds, to develop social organization. Bison, antelope, deer, horses, all of these grazing animals tend to form aggregations. Even the rodents—the prairie dogs of the great plains of the United States, for instance—tend to form social groupings. I suppose this can be explained as a protective device since the herd, in this open country, is safer than the individual alone. Some individuals can serve as sentinels while the others are eating, and any individual who becomes aware of a danger can spread the alarm to the whole group. Social organization, like fleetness and keen vision, can thus be explained as an adaptation to life in open country. The predators, too, like the wolves and coyotes, find it advantageous to act in organized groups.

There are, of course, many solitary animals in open country—the rhinoceros is notably antisocial—and, conversely, many social aggregations in the forest. Monkeys, peccaries, coatis, many animals of the tropical forest, are social. But I still think that herding is, in general, more characteristic of open country than of the forest, though this might be difficult to prove by count of individuals or species. This has a special interest in relation to the problem of explaining human origins. Man has certainly adapted to the grasslands from very early times. A single man, alone, is particularly helpless in open country—only cooperating groups of men can deal with the wary game. Of course, there are traces of the forest background in man, too, and it often seems easiest to explain man as a product of the forest margin, of the border zone between the grassland and the heavy woods. But I want to come back to this in a later chapter.

Deserts, grasslands, woodlands—these, like tropical forests, like lakes and rivers, like all of the diverse situations in the seas, are examples of different kinds of biological communities.

But what, anyway, do we mean by a biological community? And how can we analyze the activities and relationships within such communities? Before looking at these questions, I think it would be useful to consider the different kinds of units that are used in biological description and analysis—to look at the building blocks of the biological structure.

9. The Units of Life

Definition is ordinarily supposed to produce clarity in thinking. It is not generally recognized that the more we define our terms the less descriptive they become and the more difficulty we have in using them.

—THURMAN ARNOLD, in *The Folklore of Capitalism*

A friend—a scientist, but not a biologist—once remarked to me, "The trouble with you biologists is that you have become enamored with the cell."

I think I raised an eyebrow.

"Look," he said, "you always start your textbooks with a discussion of the cell. There is a diagram, with nucleus and mitachondria and all sorts of strange things carefully labeled. No one without a microscope has ever seen a cell; it is completely strange and difficult to understand. The student is lost at the very beginning of his study. Why not start with a dog or rabbit or oak tree, which is familiar, and gradually work down to the cell?"

This was some years ago. We were members of a committee invited by a university to look into its teaching program in biology. I don't think we had much effect on the university, but our discussions had a considerable effect on me. I have never tried to teach introductory biology, but if I ever do, I think I'll start with a rabbit sitting under a raspberry bush and from this gradually go into the mechanics of the situation. Neither the rabbit nor the raspberry can be understood, as functioning organisms, without reference to the cells of which they are built. But surely you don't have to start with the cells—and as I pointed out at the beginning of this book, I don't think cells are *the* basic unit of living processes.

Actually, of course, there are as many different kinds of units useful in biological study as there are points of view for such study. The units of the systematists (the people who classify organisms) are species, genera, families and the like. The geneticists deal with genes, chromosomes, mutations—a whole catalog of special units. For cytologists (the people who study cells) there is no question that the cell is the basic unit. Biochemists, on the other hand, are concerned mostly with molecules and molecular behavior.

With the natural history aspects of biology—with "skin-out" biology—we are clearly most concerned with individuals and with the ways in which these individuals interact, with their patterns of relationships. We can build up a sequence of categories, grouping individuals of the same kind into populations, grouping the populations in turn into more or less definable biological communities. The multitudinous biological communities again can be classified into a series of major landscape (or seascape) types, such as we have described. These the ecologist would call biomes—rain forests, deciduous forests, taiga, tundra, the ocean depths, shallow tropical seas, and the like. And all of these together form the interlocking system of the biosphere, the world of life.

For ecological study, our most important units are individuals, populations and communities. Population is a word all of us know and use, but like many such words, it turns out to be not easy to define. There are all sorts of definitions: not only general ones in dictionaries, but special ones in books on the biological and social sciences. I have always liked a definition that I once heard given by the economist Kenneth Boulding. He said that a population was "an aggregation of similar items enclosed by a picket fence of definition, with an entrance by way of birth and an exit by way of death." There could also, I suppose, be a certain amount of fence jumping—migration.

Economists like Boulding want to use the population idea for all sorts of things—automobiles, bolts, school children, anything that has definite characteristics and that can be counted. "Birth" would then include manufacture, "death," obsolescence. It's somewhat easier if our "aggregations of similar items" are always living individuals: at least I have less difficulty thinking about a population of goats than I do in thinking about a population of lawn mowers. The word originally referred only to human beings—the number of inhabitants of a particular city, region or country—and the whole idea is more easily extended to other kinds of living beings than it is to inanimate items.

In biology, the idea of population is closely related to the idea of species. A species, of course, is a particular kind of organism —a cottontail rabbit or a white oak. But we can't imagine a species as a single individual—it is always a group, or population, of similar individuals. A species is ordinarily defined, nowadays, as a population of similar individuals that actually or potentially interbreed, and that are separated from other populations by barriers to breeding. This, of course, would apply only to sexually reproducing organisms, but the vast majority of the organisms that we know show sexual behavior in some form and under some circumstances.

We have to say "actually or potentially" because an individual in Cuba is hardly able to mate with an individual in Jamaica, but the two individuals are classed as the same species if we judge that they could mate if they lived in the same place. There are a lot of "ifs" here—the so-called "species problem" in biology is full of ifs. The arguments and discussions sometimes give the impression of endless quibbling. But the discussion is important because a basic question in evolution involves the origin of species—and one can hardly hope to tackle the problem of the origin of species without first having a clear idea of what a species is.

We ordinarily recognize species by their appearance: the shape and color of a flower, the arrangement of spots on the wings of a butterfly. These identifying characteristics are shared by all individuals of a given species because of their common heredity. Any new characteristic that arose within a species, if favorable, would spread eventually through a whole population through interbreeding; it would be prevented from spreading to other species populations because of the barriers to breeding. A given species population thus has a common "pool" of hereditary materials.

We have now found many cases in which independent breeding populations occur in nature without any obvious differences in appearance to mark them. These have been called cryptic (hidden) species. One case that I worked on for several years involved a group of anopheline mosquitoes of southern Europe. The chief vector of malaria in that region was thought to be a mosquito called *Anopheles maculipennis*. Wherever there was malaria, this mosquito was found. But in some places where the mosquito was extremely abundant, there was no malaria, which led to intensive study of this phenomenon of "anophelism without malaria."

It was early thought that there might be two "biological races" of the mosquito: one, for some reason, capable of transmitting malaria, the other not, and a search was started for means by which these "races" might be recognized. It was presently discovered that, even though the mosquitoes looked the same everywhere, different "varieties" could be distinguished by microscopic examination of the patterns of spots on the eggs. By breeding mosquitoes of known egg type in the laboratory, it was possible to make comparative studies of different aspects of their behavior.

After a clue had been found in the egg differences, a survey was carried out over southern and eastern Europe to find out what kind of eggs were laid by mosquitoes in the different places. More kinds of eggs were found in the tiny kingdom of Albania than anywhere else. Since Albania also had a serious malaria problem, this seemed a logical place to establish a laboratory. The work was financed by the Rockefeller Foundation, and I was in charge of the laboratory from 1935 to 1939, when mosquito studies were overwhelmed by the march of events in Europe.

It turned out that the mosquito "species," *Anopheles maculipennis,* really consisted of seven independent populations which differed from each other in many fundamental respects—but which happened to look alike as adult mosquitoes. The different populations had different biting habits—only those kinds that liked to bite man could transmit malaria. Many mosquito species, believe it or not, refuse to bite man under any circumstances; others will bite him only reluctantly, in absence of some more favored source of blood. Where there was "anophelism without malaria" in Europe it developed that the mosquito populations present were kinds that preferred biting cattle or pigs or goats.

These mosquitoes differed from each other in all sorts of ways. The larvae had different habits: two of the populations were found mostly near the coast, where the larvae lived in brackish water. Another was found only in the vicinity of inland marshes, or breeding in the thick reeds of lake margins. Another preferred to breed in small ponds or stagnant ditches.

But most strikingly, each kind had characteristic mating habits. In order to maintain the mosquitoes from generation to generation in the laboratory, we had to induce the mosquitoes to mate, and this turned out to be difficult. Mosquitoes raised from one kind of egg, the variety called *atroparvus,* would mate easily enough in small cages; but none of the others would. In nature, sexually excited male mosquitoes gather in dancing swarms, to which the females are attracted. We had to find some way of getting the males to form swarms under laboratory conditions, which turned out not to be easy. We never did succeed with two kinds, and each of the others needed somewhat different conditions of light and space before the males would become sexually excited. In short, each of these mosquito populations had a different kind of sexual behavior, which would keep them from cross-mating in nature.

We tried crossing the *atroparvus* males, which would mate readily, with females of the other kinds, and we found sterility barriers in every case. With one cross, we would get eggs which would hatch, but the larvae all died in a day or two. With another cross, only a few males would reach maturity and these, when we dissected them, turned out to have their sex organs undeveloped. One hybrid turned out like the mule: we got fine, big, husky mosquitoes of both sexes, but they were completely sterile. It thus seemed that the mosquitoes were prevented from cross-mating in nature by their different habits; but that even if such cross-mating occurred, there would be no continuing offspring, no blending between the populations. Each of these kinds of mosquitoes was a reproductively isolated population, a *species*. I started out on this study hoping to find out something about the origin of species, hoping to find populations in some halfway stage toward independence. But in the end it seemed that these were fully developed species, not halfway stages. The adults all looked alike to an entomologist, but the mosquitoes could tell each other apart easily enough—and this, from the biological point of view, was the important thing.

I think we still have a great deal to learn about this problem of the origin of species, of the origin of reproductively isolated populations. How did these mosquito populations first get separated? The easiest kind of separation to imagine is geographical. A mosquito in Albania has no chance to mate with one in Spain, and one can see how the Albanian population might gradually come to differ from the Spanish population, to the extent that when an Albanian male did happen to meet a Spanish female, they would not recognize each other. Ordinarily, however, the Albanian population is in indirect contact with the Spanish population—through the mosquitoes of Yugoslavia, Italy and France—so that any changes would spread from one to another. Where populations are separated by barriers difficult to pass, like sea, mountain or desert, one can easily see how they might follow quite different evolutionary trends to the point where, if contact were again established, there would be no blending.

Human races show the process of geographical variation nicely. There are all sorts of ways of classifying races, but whatever the classification, it is clearly dealing with geographical populations. The so-called Caucasians inhabited Europe, and the neighboring parts of Asia and Africa, before they started spreading widely in modern times. The Mongolians lived in eastern Asia and, if the American Indians are classified in the same racial group, in America. The Negroes inhabited Africa south of the Sahara and the Australians, Australia. There was (and is) a blending of populations in the border zones, so that one can never say precisely where one race ends and another begins; and there have always been great movements of human population with consequent mixings. The heterogeneous population of the Indian peninsula, for instance, is explained by repeated migrations of different populations into the area. If we look at smaller racial divisions, like Celts, Nordics and Mediterraneans in Europe, we also find geographical differentiation, along with blending in contiguous areas and mixing from migratory shifts.

The recognizable varieties of men scattered over the face of the earth can be explained only in terms of a tendency for geographically-separated populations to become different, countered by a tendency for these populations to mix as they come in contact. All of the men that we know belong to one species, because they interbreed freely wherever they come in contact—such barriers as exist, like the caste system of India, or the segregation systems of the United States and South Africa, are cultural rather than biological. If some population, like the aboriginal Australian, had remained isolated for a hundred thousand years or so, we can imagine that it might have become different enough so that interbreeding would not occur—and then we would have two species.

This tendency to geographical variation is shown by all sorts of animals and plants, and some biologists insist that it must be the only mechanism for the origin of species, for the development of reproductively-isolated populations. It is hard to believe this, though, of microbes, of the oceanic plankton, or of any organisms with wide means of dispersal. And it would take an awful lot of separating and reuniting to explain the millions of species of the world. Which is why I think we still have a great deal to learn about the process of species formation.

But I am trying here to discuss the units, the categories, of biology, not the processes of evolution. Populations—species—can be dealt with in two quite different ways. I suppose we could call one taxonomic, the other ecological.

The taxonomist is interested in classifying the diverse kinds of organisms found in the world, in working out their evolutionary histories, in making a sensible and usable catalog of nature. He groups species into genera, these into families, orders, classes, phyla. By the system universally used, the genus and species names become the technical term for the population, for the kind of organism—*Homo sapiens*. He files man in a separate family, the Hominidae, along with various human fossils; puts these in the order of primates, the class of mammals and the phylum of chordates. He has built up, by this system, a series of terms that must be used by all biologists to indicate what they are talking about. It is a highly abstract system, but it starts with species—with populations.

The ecologist starts with these same populations, but he is interested in how they live together. He regards his species, not as members of genera and families, but as parts of biological communities. When he starts out to try to analyze and describe these communities and to work out an orderly system whereby communities can be classified and compared, he runs into all sorts of difficulties. The taxonomists have problems with their genera, families and orders—but the taxonomists, at least, have reached a general agreement about their systems and their rules. The ecologists, it seems to me, are still lost in the biological communities that they are trying to analyze. But the theory of the biotic community should, I think, be the subject of a separate chapter.

10. The Biological Community

Food is the burning question in animal society, and the whole structure and activities of the community are dependent upon questions of food-supply.

—CHARLES ELTON, in *Animal Ecology*

Community is a common English word; we all know what it means without looking it up in the dictionary. It implies a group of people living and working together, mutually interdependent in all sorts of obvious and subtle ways. We may rebel about "community spirit" and sneer at "community leaders" because we are trying to hold onto our individuality, our privacy. But the individual's rebellion is simply more evidence of the pervasiveness of the community ties. The individual may get private satisfaction from his rebellion without doing public damage—if public damage is apparent, he gets locked up—but the rebellion is doomed to frustration because no man can live by himself.

Similarly, no organism can live by itself—at least no kind of organism does live by itself. If I try to imagine a hypothetical case, I am driven to obscure and far-fetched possibilities. Animals cannot be considered because they directly or indirectly depend on the green plants, the plants with chlorophyl, for their food. But these green plants are not independent, either. With energy from sunlight, they can build up sugars from carbon dioxide and water, but they cannot live on sugar alone. For their other substances they must depend on other organisms. Nitrogen, for instance, is needed by all living things, but only a few organisms—certain bacteria and algae—can utilize free nitrogen from the atmosphere, and the rest of the living world is dependent on these to get their nitrogen supply. Maybe some of the algae, or some of the odd bacteria like the ones that are able to get energy from sulphur, could live alone, but it is hard to imagine.

The problem is similar to imagining a man alone. It can be done—we have Robinson Crusoe. But even so it is a temporary aloneness, made possible by materials and skills carried over from previous associations with other men. And Crusoe was much happier after he found Friday.

Interdependence thus underlies the community idea both with man and with nature. But there are very important differences. The human community—the social community—is made up of individuals of the same species, made up of people. The biological community is made up of many different kinds of organisms—plants, animals, microbes. The social community is a grouping within a population; the biological community is a grouping of many different populations. In comparing the two, then, we are dealing with analogies, and while analogies form powerful tools of human thought they can, like other tools, be dangerous if misused.

The gap between the human social community and the biological community has grown with the increasing complexity of culture, with the development of civilization. With food-gathering peoples or with primitive agriculturists—with Australian tribes or Amazonian Indians—the difference between the social and the biological community is clear, but the social community can be looked at as a part of a particular biological community. In studying the social community we are interested in beliefs, in marriage customs, in institutions of various sorts. We are also interested in ways of getting food, which leads us directly into the biological environment, and the economy of our tribe will differ depending on whether the tribe inhabits rain forest, grassland, desert, seacoast, or what. But with the increase in man's control over nature, with the development of communication, transportation and exchange, the relations between the social community and any particular biological community become more and more diffuse and indirect.

The town of Ann Arbor, Michigan, where I live, can be studied as a social community, or as a series of social communities. In looking at it from this point of view, we are interested in the division of skills and services, in the habits and customs of the people, in their institutions—schools, churches, clubs, jails. We can examine sub-groupings; argue, for instance, about the degree of separation of the people associated with the university from the other people of the town. But we find little connection between the social community of Ann Arbor and any particular biological community.

Ann Arbor, to be sure, represents a biological aggregation. We have people, dogs, cats, rats, mice; and cockroaches in the kitchens, though probably no bedbugs in the bedrooms. Chickens and cows are not allowed within the city limits; I'm not sure about horses, though I have seen none housed within the town. There are shade trees and lawns and gardens, and occasional untidy vacant lots. Squirrels, raccoons, opossums and many kinds of birds have adapted to these conditions and get along very well. But where is the biological unity in this?

If we start with man and look at his food habits—taking food relations to be the cement that binds the biological community together—we find that he lives on wheat from Nebraska, cattle from Texas, coffee from Brazil, sugar from Cuba, fish from Florida or Maine and so on. Ann Arbor is physically located in what was once deciduous forest, but this has left almost no trace in the culture and institutions of the people. Ann Arbor as a social community hardly differs from similar communities located in the prairies or on the seacoast. The fact that it is a university town is far more important than the fact that it is located in countryside once covered with deciduous forest. But this is a consequence of the very special relationships that have developed between man and other organisms with the evolution of his culture, his civilization.

To avoid the danger of confusing the different concepts of biological and social community, ecologists often use the word *biocenosis* for the biological community. This seems to me to illustrate nicely one of the dilemmas of technical vocabulary. We have the alternatives of taking an ordinary English word and giving it a special meaning, or of inventing a new word free from the ambiguities and connotations that accumulate around any word with a long history in the language. Ann Arbor might, mistakenly, be thought to be a biological community, but there is no danger that it will be thought to be a biocenosis.

But if we use biocenosis, we lose the implied analogy with the interdependence of the butcher, the baker and the candle-stick maker, and the analogy seems to me more helpful than harmful. The danger lies not in the analogy, but in the possibility of confusing analogy with identity. There is an equivalent danger, I suspect, in substituting a strange word for a familiar word in these cases: we are liable to fool ourselves into thinking we have produced a new thought, when we have only produced a new word, and we get a misleading feeling of being "scientific" and precise. This is harmless enough in isolated cases, but it can become habit-forming—and it has become a dangerous habit in ecology. Biocenosis leads easily to biomes and biochores, to ecosystems, ecotones and seres. These are all lovely words, but they don't really say anything new. The trouble is that the word-coiner, sinking blissfully into his addiction, gradually loses all communication with the outer world. He emerges from time to time to complain that the world doesn't really understand or appreciate his important thoughts—meaning his big words. New words get general acceptance fast enough when they really correspond with new ideas or new discoveries —like the genes and chromosomes of the geneticists, or the television and radar of everyday speech.

I prefer, then, to stick to community. But I have trouble

whether I write about communities or about biocenoses, because I can't give a neat or precise definition in either case. We use the word community to express the interdependence of organisms. But where does one community stop and another begin? How do we recognize a natural community when we find one?

The community idea simply can't be given a rigid definition. The difficulties are the same as those that apply at every stage in attempts to subdivide and classify the parts of the biosphere. There is trouble with our major division into seas, inland water and land. We have more trouble with the next step of subdivision into landscape types—the *biomes* of the ecologists—trouble in setting precise limits to the concepts of desert, grassland, scrub, forest. We have even more trouble when we try to separate out particular kinds of communities within the deserts, the forests, the lakes or the seas.

I like to think of a community as an aggregation of different kinds of organisms interdependent on one another in all sorts of ways, but relatively independent of outside organic influences. A rotting log in a forest teems with life of all sorts, and the organisms within the log are mutually interdependent in many ways. But I would not say that the inhabitants of the log form a community because their activities cannot be understood without constant reference to the forest environment in which the log occurs. The log is a part of the larger forest community. There may, however, be many different forest communities in a given region, each of which can be studied and understood with little reference to the others.

In northern forests, particularly, communities can often be defined in terms of some dominant species of tree. We can recognize pine forest communities, oak-hickory forests, maple forests, spruce forests and the like. The whole character of the community may in a real sense be controlled and dominated by a single common species of plant. This has led to the idea that there must be, in every community, some dominant and characteristic organism. But this idea breaks down rapidly in tropical forests, and in the north it doesn't work very well in lakes or in the seas.

However we define our communities we are bound to have zones of transition. Occasionally these are so abrupt that a line can be drawn, as when the margin of a river or a lake meets a forest. But more often they are gradual, one type of vegetation with its associated animals almost imperceptibly giving way to another as conditions of climate or soil change. The changes can be measured by making counts of the changing numbers of characteristic plants or animals, but there is no sharp boundary between one community and the next and often there is little to be gained by arbitrary attempts to create boundaries. Nature has not been arranged in accordance with the neat grid of any surveyor. Sometimes it is useful to impose a grid, but it is always dangerous when we forget that the grid is arbitrary.

Species, populations, kinds of organisms, are best defined in terms of reproduction—we are asking who mates with whom. Communities or mutually interdependent aggregations of populations are best defined in terms of food relationships—we are asking who eats whom. There are, of course, all sorts of other interrelations—protection, support, transportation. There may even be interdependence among species in reproduction, as with the flowering plants that depend on different kinds of insects for pollination. But food relations form the basic cement that binds the community together: in studying the structure of the community, we soon find that we are basically concerned with the endlessly diverse ways, some obvious and some subtle, in which organisms depend on each other for food.

"Why do we have to eat?" a friend once asked me.

It seemed a silly question, and I remember feeling quite impatient. "You have to eat to live. Why do you put fuel in your automobile? It has to have gasoline to run—some source of energy. If you look at man as a machine, the motor is running all the time; it's only cut off with death. Every organism has to have some regular source of food."

He kept pushing me about this energy business. Animals eat

animals until, sooner or later, we come to animals that eat green plants. And green plants? Somehow, through the agency of chlorophyl, they are able to use the energy of sunlight to build up sugars from water and carbon dioxide.

"Oh," he remarked with sudden illumination, "in eating you are getting your share of the sun."

We both liked that, and pushed the idea around for a while. "Well," he said, finally satisfied, "I get my share of the sun in a supermarket," and walked off.

But there are a lot of stages between the sun and the supermarket. Man has manipulated these stages more and more for his own convenience, but he still has not been able to free himself from the sun or from the intervention of chlorophyl in green plants. Man has been able to modify the appearance of the links in his food chain in all sorts of elegant ways, but he is still chained by the laws of nature, by the energy sequence of the biotic community.

If we look at any community, we can group the organisms that make it up into three broad categories: the producers, the consumers, and the decomposers.

The producers are the green plants, the organisms with chlorophyl. By using the sun's energy to build up sugars through the process called photosynthesis, these plants accumulate the fuel that keeps the whole living world going. All of the energy of life comes from the oxidation—the burning—of these sugars, which go through an endless diversity of chemical sequences before they finally end again as water and carbon dioxide. The plants use some of their own sugars for living and growing, but they produce more than they need and the rest of the organic world depends on this surplus.

This carbon cycle provides the basic motive power for life as we know it, but a wide variety of substances besides carbon, hydrogen and oxygen are involved in the chemistry of living and each of these substances is used over and over again by organisms, making endless patterns of interlocking cycles. The nitrogen cycle is particularly important because, while nitrogen is the commonest gas in the atmosphere, most organisms cannot use it in this pure form. Even the green plants, then, must depend on nitrogen-fixing bacteria for their nitrogen—and the bacteria, in turn, must depend on the plants for their carbon.

Animals, then, basically are consumers, living off the green plants. They give many chemicals back to the plants in excretions or from the decomposition of their dead bodies, but they live by burning the carbon compounds that they get, directly or indirectly, from the plants, and the end product of this burning, carbon dioxide, can only be recovered by the plants through photosynthesis.

The food chains—the ways of passing around the carbon supply—are endlessly diverse in any community. The patterns become so complicated that it is probably better to think of food "webs" rather than food "chains." But the chain is more easily visualized—even if oversimplified. We can think of chains like grass-grasshoppers-frogs-snakes-hawks. The grasshopper, the frog, the snake or the hawk may die from some intervening cause and its body material be returned to the soil by the decomposers. Or the grass may die and be decomposed without passing through any animal at all. The decomposers—mostly microbes and fungi—are always with us, playing their very essential role, reducing the carcasses of plants and animals to their component materials, so that the chemical cycles can start all over again.

The animals that live directly off the plants are generally very small and very numerous. The British ecologist Charles Elton has called them the "key industry animals" because all of the rest of the complicated animal economy depends on them. In the sea (and in fresh water) the producers are generally microscopic organisms, floating in the surface plankton; and the key industry animals are either microscopic or very tiny, also forming part of the drifting plankton. On land, the producers are mostly ferns and seed plants, and the key industry animals are the hordes of many kinds of insects that live directly on the leaves of these plants.

In the animal food sequence, the ecologists often write about the "pyramid of numbers." The key industry animals are generally both very small and very numerous; the animals that eat them are generally larger and fewer, and so with each step until we come to the top predators—the sharks, lions, and eagles—which are rather large and rather scarce. The total mass of organisms at each successive level in the food sequence must necessarily be considerably smaller than at the preceding level because energy is constantly being dissipated in the business of catching and eating and living, so that at the final consuming level we generally find a relatively small number of rather large animals.

But there are many kinds of exceptions to this sequence. On land, the grazing mammals live directly off the vegetation and achieve large size and the largest of all land animals, the elephant, lives directly off plants. But since the elephant represents a sort of size eddy in animal economy, I would hardly call it a "key industry animal." In the sea, the biggest things of all, the whalebone whales, short-circuit the whole food chain sequence to live directly off the tiny key industry animals of the plankton, which they filter out by means of their gigantic sieves. But the principle of relative mass would hold, since the mass of microscopic animal plankton in the sea is vastly greater than the total mass of the filtering whales.

The study of how animals get their food is endlessly fascinating. The first-order consumers, the key industry animals, the animals living directly off plants, would seem at first thought to have an easy time of it. Plants in general don't move, so they don't have to be caught, and they are apparently defenseless, lying around waiting to be gobbled up.

But the first-order consumers have their problems, the chief of which is dealing with the cellulose walls of plant cells. It is surprising, when you stop to think of it, how few kinds of animals have solved this problem. On land, we have a few kinds of molluscs (snails and slugs), various orders of insects, and the herbivorous mammals. All of the rest of the land animals are living off these, or living off special parts of plants like seeds, fruits and tubers, or living off detritus of one sort or another.

The first-order consumers must have either a special enzyme capable of breaking down cellulose, or some method of grinding up the plant cells to make a digestible pulp, or a symbiotic relationship with a microbe capable of dealing with cellulose. Most herbivorous animals combine at least two of these three methods. Cows and other grazing mammals, for instance, have teeth well adapted for grinding up leaves; but they also depend on special flagellate protozoans which live symbiotically in their digestive tracts, and which break down the cellulose.

Symbiotic—mutually dependent—relationships between quite different kinds of organisms are very common and widespread in nature, and often turn on food-getting or digestion. The lichens form the classic case. Each "species" of lichen is made up of a fungus and an alga growing in a close, symbiotic, relationship, and neither partner in the union is ever found alone in nature. The fungus provides the structure in which the single-celled algae grow and absorb water and salts from the environment. The algae, through photosynthesis, build up organic food for both partners.

Both fungi and algae have commonly developed symbiotic relations with other kinds of organisms as well as with each other. Coral animals always contain symbiotic algae and the strikingly colored mantles of the giant clams of the Pacific (*Tridacna*) form an elaborate arrangement for maintaining symbiotic algae. The clams are involved in a sort of intimate greenhouse operation for growing their own vegetables inside themselves. A great many different sorts of seed plants have fungi growing in a symbiotic relationship with their root systems since the fungi are apparently more efficient than the plant roots in gathering needed inorganic salts, and the fungus in turn gets organic food from the seed plant.

The word *symbiosis* is sometimes restricted to these intimate reciprocal relationships. But sometimes—and more properly—it is used in a very broad sense to cover all close relationships between two different kinds of organisms. The word essentially means "living together." There are all sorts of ways in which organisms live together, and it turns out not to be easy to classify these ways.

An orchid growing on a branch of a tree is living in close association with the tree. As far as we can tell, the tree gets neither benefit nor harm from the orchid, but the orchid definitely gets the benefit of a place to grow, a perch. This sort of relationship—benefit to one partner and neither benefit nor harm to the other—is often called *commensalism*. Barnacles growing on a whale would be another example—the barnacles get transportation. Cockroaches in the kitchen could be regarded as commensal, since it is difficult to demonstrate that they cause either harm or benefit to the housewife though the cockroaches get the benefit of a warm place to live and easily available food scraps. But the cockroaches may annoy the housewife and there is a nice gradation between the annoyance of cockroaches for the housewife and the annoyance of fleas for a dog. The alleged benefit dogs get from fleas is very questionable, and they clearly are living directly at his expense, so we call this relation parasitism. But where both partners clearly benefit, as with the fungus and the alga in the lichen, we call the relationship *mutualism*.

There are many difficulties in a system of classification like this. In the first place, it is often hard to determine benefit or harm. And even where this can be measured with some certainty, every possible gradation seems to exist, so that we have the old problem of where to draw lines of definition in gradually changing series. The *Plasmodium* that lives in the red blood cells of man and causes the disease called malaria is clearly causing immediate harm to the individual in whose blood it lives. The malaria parasites cannot live without man, though man can live easily enough without malaria parasites. The protozoans that live in the intestines of termites, where they digest the cellulose eaten by the termites, on the other hand, are clearly of direct benefit—the termites cannot live without their protozoans, and no one has ever been able to keep these particular protozoans alive outside of the intestinal tract of their termite hosts. But in many cases where we find micro-organisms associated with an animal host, we can't be sure whether they are benefiting or harming the host, or whether they have any effect at all; and sometimes the micro-organisms may be beneficial under some circumstances and harmful under others.

The relations between insects and plants illustrate this difficulty. A great many insects eat plants—they are parasites on the plants. But often the same species of insect that in the larval stage eats the plant, serves as an adult to pollinate the plant; plants, for reproduction, are often dependent on insects that in turn are dependent on the plants for nutrition. Generally the reciprocity is rather loose and general—caterpillars feed on leaves, and the adult butterflies and moths pollinate flowers, but usually a particular species of butterfly is not important for the pollination of the flowers of the plant on which its caterpillar lives.

Sometimes, however, close reciprocal relationships are developed. In the American southwest, for instance, the flowers of the yucca plants (Spanish bayonets) are pollinated by a particular kind of moth (*Pronuba*) which has a special pollen-gathering apparatus for collecting a little ball of sticky pollen from the yucca flowers. When the eggs of the moth are mature, it pierces the ovary of another yucca flower to lay an egg inside, then climbs to the top of the pistil to rub some of the pollen on the open stigmatic tubes. The yucca flowers can be pollinated only by this moth, and the larva of the moth can develop only in the yucca seed pods. The moth larva is parasitic on the yucca —though the moth lays so few eggs that the plant is able to develop an adequate supply of seeds. And the moth pays for its food by insuring the cross-fertilization of the yucca flowers. Is this then parasitism or mutualism?

It thus often becomes difficult to distinguish between parasitism and mutualism. But it is also difficult to distinguish between parasitism and predation. These are both common words and we all think we know what they mean: parasitism involves lots of little things living off big things, fleas on a dog or worms in a sheep. Predatism involves big things killing and eating other things, usually smaller—foxes eating rabbits, or lions eating gazelles.

The extreme cases are different enough, but when we start out to make precise definitions of our words, or to classify cases, the differences start to blur. We are dealing with second, third or fourth-order consumers, and we soon find ourselves lost in the complicated tangle of different ways in which animals live off each other.

We may say that the predator kills the prey immediately, while the parasite either manages to live off the host without killing it, or kills it slowly. We can certainly sort out a series of categories in this way: death immediate and certain; death postponed but certain; death probable but not inevitable; death unusual; and finally, one kind of animal living off another with death never a consequence. But here we have not two contrasting terms, parasite and predator, but a series of five categories, the last three rather arbitrarily separated. And if we start sorting things out according to these categories, we get some strange bedfellows from the point of view of our usual ways of thinking.

Death immediate but certain covers our usual idea of predators—lions, weasels, crocodiles, barracuda, octopuses, dragonflies, and starfish prying open oysters. We might get in some trouble deciding what we meant by immediate. The starfish, for instance, takes a little time to pry open the oyster and extrude its stomach into the oyster shell to digest its living prey. The cat may play with the mouse for quite a little while, and the spider may wait for its victim to tire from the hopeless struggle in its silken trap. But this is quibbling.

Death postponed but certain involves cases that are sometimes called "parasitism" and sometimes "predation." The predatory wasps, for instance, hunt their spiders or caterpillars or whatever prey they specialize on, and immobilize it by stinging—but they don't kill it. The prey is carted off and stored in a cell or burrow and sealed up with a wasp egg, to be eaten at leisure by the developing wasp larva, a sort of deep-freeze arrangement.

The so-called parasitic insects also belong in this category. In this case the female of the parasite lays her eggs in another insect, and the larvae develop but avoid eating vital parts of the host until the end—but in the end the host is always completely devoured. There is no evolutionary tendency, as in other parasitic relations, to avoid killing the host. Micro-organisms which inevitably kill their host would also belong in this class, but the only instance I can think of at the moment is rabies virus, which always causes death once the virus has become established in nerve tissue. With most parasitic micro-organisms, the likelihood of death of the host varies from probable to unlikely, depending on the particular species of parasite and host involved. And this is true with most of the relationships commonly recognized as parasitic.

If we take a long-term view, the distinction between beneficial and harmful becomes even more difficult. In general, the reproductive rate of any particular kind of animal is adjusted to the hazards of existence of that kind of animal: elephants, which have few hazards, have a low reproductive rate; oysters, which must face many hazards, produce millions of young. Parasites and predators are prominent among the hazards of existence for most animals, and reproductive rates are adjusted to these hazards. When hazards are removed, the balance in the community may get upset, as with the deer of the Kaibab Plateau in Arizona. Were the predators, there, harmful or beneficial?

The "balance of nature" is never completely stable; it is always teetering in some way so that animals vary in abundance from year to year or millennium to millennium. The unbalance sometimes becomes great enough to lead to the extinction of some kind of animal—witness the dinosaurs, the pterodactyls, the saber-toothed tigers and mammoths. We do not really know, in the case of animal types that became extinct in the geological past, whether the extinction was caused by climate changes, by competition, by new predators or parasites, or what. In the cases of extinction within historic times, man always seems to have been the primary agent, sometimes directly as a predator, sometimes indirectly through changing the landscape or through accidentally or purposefully shifting organisms about from continent to continent.

Man as an agent of change in the biotic community is a new phenomenon. But man is still a part of nature, and the changes he brings about must be in accord with natural principles—though this may turn in part on how you define "natural." Certainly human operations have vastly altered the balance of nature and drastically changed community relationships. This, from many points of view, is deplorable—but it also provides gigantic experiments in the alteration of biological communities from which we can learn a great deal about the community structure.

Disease is one aspect of the parasitic relations within communities, and disease patterns have been greatly altered by human activities. But the problem of disease seems to me interesting enough to warrant exploration in a separate chapter.

11. The Natural History of Disease

. . . . soldiers have rarely won wars. They more often mop up after the barrage of epidemics.

—HANS ZINSSER, in *Rats, Lice and History*

The biological community, then, is a vast and complicated system for sharing and distributing the energy of the sun among a diversity of life forms. One of the ways in which organisms get their share of the sun is called parasitism: living off another organism in an intimate association that, ideally, causes a minimum of immediate harm to the victim. But some immediate harm is always caused, by definition. If the relation were not one-sidedly harmful we would call it mutualism instead of parasitism. When the harm that is caused is obvious, we speak of disease; and the parasite causing the disease is called a pathogen, from the Greek *pathos* and *gens*—a bearer of suffering.

This immediately implies a point of view. We are not looking at the world from any impartial, cosmic position, we are looking at it from the point of view of the host, the victim, the supporter of the parasite. We are, not surprisingly, being human because disease is an important human problem.

Man has always been bothered by disease. At least comment on disease goes back into the oldest records and every culture that we know, however primitive, has some way of dealing with disease. The history of science—and of biology in particular—is completely entangled with the history of medicine, of ways of dealing with disease.

I suppose the demon theory of disease is the oldest, and ties in with an animistic view of nature which sees spirits everywhere and hardly distinguishes between animate and inanimate, man and animals, rocks, winds and clouds. Sickness is caused by a demon getting into a man; cure can then be effected by getting the demon out. This becomes the function of the witch doctor, and the sorcerer; these accumulate a special store of knowledge which is passed on to chosen initiates. I sometimes like to think that all of our learning, all of the special lore and ritual of scholarship, descends in a straight line from the accumulated wisdom—and mumbo jumbo—of the witch doctors of our ancestral tribes.

This brings up the problem of magic and science, which I would just as soon avoid. But it is interesting that while our ancestors were chasing demons, they were also ransacking the natural world for drugs of one sort or another and coming up with a remarkable collection—to which civilized man, with all of his science, added very little until the recent discovery of sulphas and antibiotics. How primitive man discovered the ways of extracting, preparing and using all of these drugs, poisons and foods, remains one of the great mysteries of human prehistory. Of course many of the herbs and remedies had no helpful physiological effect—however potent they may have been psychologically. But we are coming more and more to realize that others did have specifically helpful effects, and pharmacologists have developed a lively interest in the remedies of folk cultures.

But this is no place to review the history of medicine or of man's attitude toward disease. I want to look at parasitism, at disease, as a biological phenomenon; as an example of community relations. This means that I am concerned with one kind of disease—infectious disease, disease caused by germs. Germs are the demons of our modern world.

Leeuwenhoek discovered the world of subvisible living things toward the end of the Seventeenth Century, but the understanding of the roles of these microbes in the economy of nature did not begin to emerge until about a hundred and fifty years later. It took a long time to realize that microbes followed the same rules as other kinds of life, that they could be grouped into species and families and classes; that they reproduced as other organisms do; that each kind had its characteristic requirements for survival and dispersal. It took a long time, in other words, to build up information about the natural history of microbes. There was no traditional knowledge to build on, and the gathering of knowledge had to depend on the development of special methods and instruments, especially on the improvement of the microscope.

We are apt to date our understanding of the role of microbes in disease from Louis Pasteur, just as we are apt to date our understanding of evolution from Pasteur's contemporary, Darwin. We can trace ideas of infection before Pasteur, just as we can trace ideas of evolution before Darwin, but the two men do strikingly and truthfully serve as markers for great shifts in the direction of scientific thought. Pasteur's greatest contribution, perhaps, was killing once and for all the idea that microbes arose or could arise by spontaneous generation. Once the fact had been firmly established that microbes, like all other organisms, came only from other microbes of the same kind, the problems of their natural history, of their behavior, could be studied intelligently.

We can hardly realize now how difficult it was to kill the idea of spontaneous generation. I first became fully aware of this when I read a book by Robert Chambers, published in 1844, *Vestiges of the Natural History of Creation.* It is a little-known book now, but of considerable importance in the history of evolutionary thought. Chambers was a thorough convert to the idea of evolution, and he wrote with an easy and persuasive style that won him a wide audience. It can be seriously argued that the widespread discussion of Chambers' book in England was an important factor in the ready acceptance of Darwin's ideas in 1859.

In reading this book, I found it hard to remember that it was published in 1844—until I came upon the discussion of spontaneous generation. Chambers, like Lamarck, believed that living things were constantly being formed from non-living materials. And indeed, if life started once, spontaneously, according to natural laws, why shouldn't the starting continue, according to these same natural laws? Life starting only once had the aspect of the miraculous, of divine interference.

In proof of the continuing new formation of organisms, Chambers cited not only the microbes that appeared in their teeming abundance whenever conditions were appropriate, but also the worms that appeared, for instance, in the muscles of pigs. How else could these worms have got there if they had not been formed, spontaneously, out of the surrounding muscle? Until this possibility had been absolutely rejected, there was no motive for the hard detective work needed to untangle the devious and complicated life histories of the parasitic worms. Now, when we find a parasite in some part of a host's body, we ask how it got there—knowing that it must have got there somehow, that it could not have arisen spontaneously. But how difficult it was to arrive at this point of view, so obvious and commonplace to us!

Pasteur, with his ingenious and dramatic experiments, demolished the idea of spontaneous generation. He also firmly established in the public mind the idea of infection, of microbes as causative agents of disease. In the approximate period between 1870 and the turn of the century, there was a frantic search for the kind of germ that might be the cause of each kind of disease, and most of the important human pathogens (aside from the viruses) were discovered. Robert Koch, perhaps the greatest of these germ hunters, worked out a set of rules for the game, a statement of the conditions that must be met before a particular microbe could be accepted as the causative agent of a particular disease. The rules were needed when it became clear that microbes were everywhere, associated with health as well as disease, and that the discovery of a germ in a patient with a disease, was not in itself proof that the germ was causing the disease.

Man likes to simplify things, to find single causes, to find an order in nature that corresponds with an orderly arrangement of ideas in his own mind. This surely is one of the great drives of thought, leading to many of the great ideas of philosophy, religion and science. But nature is also frightfully complex, perhaps too complex ever to be "understood" through the processes of our limited brains—and our fondness for single causes has probably got us in trouble more often than it has helped us.

It would be intellectually satisfying if we could find a single cause for disease, and we keep trying. The demon theory of the primitives was a single cause. Galen's concept of the four bodily humors (melancholic, sanguine, choleric and phlegmatic) that were balanced in health and unbalanced in disease provided an elegant and satisfying theory for the Middle Ages. And then came germs. Germs demonstrably were the causes of many diseases and, naturally enough, there was a period when a germ was seen lurking behind every disease. If I may vastly oversimplify, we can look at medicine during the last half century as a gradual retreat from this position, a gradual realization of the diversity and complexity of disease situations.

Any contemporary classification of diseases includes a variety of categories, like deficiency diseases (due to some lack in the diet), mental diseases (with a whole catalog of psychoses and neuroses), hereditary diseases, congenital diseases, neoplastic diseases, poisonings, accidents, and so on. I don't know of any classification that isn't in part overlapping and conflicting. Even the infections, the diseases caused by germs, become more complicated as we learn more about resistance, susceptibility, immunity and environmental influences. The progress of medicine in controlling disease has been startlingly rapid in the last fifty years—but there is still so much to learn!

In looking at disease—or in looking at the ultimate consequence of disease, death—it seems to me useful to distinguish between physiological and ecological causation, between "skin-in" factors, inherent in the organization of the individual, and "skin-out" factors, arising in the environment outside of the individual. At one time, I thought I could distinguish quite clearly between these two kinds of things, but I am no longer so sure; internal and external factors interact in many disease situations, and perhaps some element of each is always present. But while the distinction is not absolute, I still think it is useful.

Before exploring this, let's look at the general question of death. Death is a consequence of the organization of life into packets, into individuals. The life stuff itself, the germ plasm, is potentially immortal and we believe has been continuous at least for the past two or three billion years on our planet. But individuals, species, organizations of the life-stuff and ways of organizing it have come and gone.

Cultures of tissues in the laboratory can be kept alive indefinitely if care is taken to provide nourishment, to remove waste products, and to cut away excessive growth to prevent overcrowding. Bits of muscle or skin or other tissues can be taken from an animal or plant and so cultured. But so far we have not been able to keep the organism itself alive indefinitely—death, then, is a property not of the life-stuff, but of the organization. A mouse or a monkey or a man, given the best of care and protected from all external hazards, grows old, shows a series of progressive changes, and presently dies. The possible life span, with each kind of organism, is characteristic, so that a mouse is old at four years, a monkey perhaps at fifteen, and a man—shall we say at 90? Each organization seems to be self-limited, to have a built-in time clock that presently runs down. With most animals and many plants, growth stops at a characteristic size, and this stage of maturity is maintained for a while and then degeneration, senility, sets in. Some organisms—particularly some fish, reptiles and trees—do not show such a sharp limitation of either size or age. But there are always exceptions.

Death from old age, from the running out of the time clock, is clearly an internal matter—physiological death. This has become of increasing interest to man as he has achieved control of ecological death, of death due to external factors. So far we have made no appreciable progress either in understanding or controlling the process of aging. Possibly we never will be able to control it, but that is no reason not to try.

Physiological death is very rare in nature. You never come across a senile rabbit or fox. It is said that senile lions take to man-eating, because men are so relatively easy to catch. But I doubt whether even a lion could survive very far into senility in his natural habitat. There is no home for aged lions in the veld, and hyenas are said not to indulge in sentiment. We think of death from old age as "natural death" and we thus come across the paradox that natural death is uncommon in nature, unnatural. I doubt whether it has ever been common with any animal except man, and with man it must be a recent development, a product of civilization, a consequence of increasing independence from the controls of the biological community. Some human cultures within historic times have not been able to afford the luxury of supporting their aged.

But the differences between internal and external causes of death are not clean-cut. We can regard death from old age as physiological, independent of external circumstances. But completely independent? The speed of living seems to have something to do with the aging process—the stresses and strains of life, products both of temperament (internal) and circumstances (external). Climate, diet, occupation, daily habits, many kinds of external factors may influence the rate of aging, the length of life.

With death at an advanced age, we can often note a specific cause, most frequently some defect in the circulatory system or the development of cancerous tissue. These categories of ills, the commonest causes of death for modern Western man, have both physiological and ecological aspects, and the problem of sorting out influences is not easy. Some of the cancers seem to be purely physiological, like one rather rare type of cancer of the large intestine which inevitably develops in individuals of a certain heredity at about the age of forty. Other cancers, like those caused by radiation or carcinogenic substances, are ecological. Most cancerous tumors probably result from a combination of both types of factors.

Infectious disease, on the other hand, would seem to be purely ecological—the consequence of the interaction of host, parasite and physical environment. The study of the ecology of infections has, in fact, been extremely rewarding. Epidemiology, the study of the incidence and transmission of disease, can logically be regarded as an ecological subject. But the reaction of different individuals to exposure to an infection, under apparently identical circumstances, will not be the same; we have the problem of individual resistance to infection, which is at least partly physiological.

Accidents—being hit by an automobile or by lightning—might be regarded as purely external, ecological. The incidence of any particular kind of accident is related to the way of life of the population concerned. The danger of being hit by an automobile is a consequence of living in one kind of environment; the danger of being hit by a falling coconut or lost in a seagoing canoe is the consequence of a different way of life. But since psychologists have come up with the idea that some people are "accident prone" it looks as though even this category of ills has an internal element.

Here I want to look at disease—and death—from the ecological point of view; I suppose the foregoing digression serves chiefly to show that I realize that the ecological picture is not the whole picture. But what is the role of disease in the economy of nature? Meaning by disease, infectious disease, parasitism.

It must be important because the parasitic way of life has been adopted by an extraordinary variety of organisms—including representatives of almost every major group (phylum) of microbes, plants and animals. The viruses, as far as we know, are all parasitic. Many different types of protozoans and bacteria have become parasitic. Algae, with their chlorophyl, are mostly found in mutualistic rather than parasitic associations, but with the fungi, parasitism as well as mutualism is common. There are even a few cases of parasitism among the higher plants—dodder and mistletoe, for instance. In the animal kingdom parasitism is rife—roundworms, flatworms, segmented worms, arthropods, molluscs and some of the minor phyla include a bewildering diversity of parasitic forms.

Even in the vertebrates there are parasitic phenomena. The European cuckoos and the American cowbirds are commonly called "parasitic" because of their habit of laying their eggs in the nests of other birds, to be raised by the foster parents. But this is a rather special form of parasitism. The sea lampreys that have caused so much damage in recent years to fish in the Great Lakes are in some ways parasitic, since they attach themselves to other fish and suck their blood. But they have not established intimate associations with their hosts, and could equally well be called predatory. They are like fleas, mosquitoes and vampire bats in this respect.

If we look at this the other way around, almost every kind of free-living organism has parasites specialized for living at its expense; there are even parasites of parasites. The community sometimes seems overloaded with parasites. I have made autopsies of hundreds of South American mammals in the course of yellow fever work, and I do not remember ever finding one that did not support a collection of parasites—fleas, ticks and mites on the outside; worms in the liver, worms in the lungs, worms in the intestines, protozoans in the blood stream. These animals seemed perfectly "healthy" when caught, but each was supporting a whole zoological garden of parasites.

In one sense, of course, the whole animal kingdom is directly or indirectly parasitic on the plant kingdom. In a narrower sense, every species of flowering plant has its special collection of parasites—insects, nematodes, fungi, all sorts of things that eat its leaves or bark, bore through its tissues, live on its roots. A particular species of plant may support hundreds of species of animal parasites. In a tropical forest one can sometimes search for a long time to find a perfect leaf—a leaf that has not been nibbled or cut or scarred by some parasite. Yet the trees of the forest look "healthy" enough.

Damage from parasites and disease is the normal, the common, situation in nature. Health, in the sense of freedom from parasitism, is unusual for an individual or for a species. On the other hand, catastrophic situations, situations in which a population is threatened with severe damage or with extinction (epidemic situations) are also unusual. It seems to me, thinking over my experience and reading, that catastrophic situations resulting from parasitism almost always turn in one way or another on human interference with the balance of the biological community. Yet catastrophic situations have, time after time, developed in relation to human activities. Epidemics, in human experience, are far from unusual. Why is this?

One could answer in general terms by saying that man, as an animal species, has embarked on an unusual enterprise, so that it is not surprising that it should involve unusual consequences. The whole course of man's cultural development turns on shifting the balance of nature, and I don't think anyone is going to propose that we go back to the prehuman state of being just another rather uncommon species of primate, living in a nicely adjusted equilibrium with the forest. But we might as well try to understand what we are doing—and try to circumvent catastrophes.

The biological community, undisturbed by external forces like man, or volcanic eruptions, or hurricanes, tends to reach and maintain a state of equilibrium. This means that the relative (as well as absolute) numbers of individuals and populations are fairly constant. This can only come about if birth rates and death rates in the various populations are balanced. One task of the ecologist, then, is to look for the means by which this balance is maintained.

Insofar as death is concerned, the ecologists distinguish between causes that they call "density-independent" and those they call "density-dependent." In the first case, the likelihood of a particular individual being killed has nothing to do with the numbers of individuals of that kind present. Such density-independent causes of death are largely physical or climatic, turning on conditions of temperature, rainfall and the like. Unquestionably a great many organisms are killed by the hazards of climate, like frosts and droughts, though ecologists are by no means in agreement about the importance of this.

In general, biological causes of death like parasitism and predation show a density-dependent relationship. That is, the death rate from a particular cause will tend to increase, relatively, as the population increases. To use a crude example, as rabbits multiply, foxes will tend more and more to concentrate on rabbits as food, but when rabbits become scarce foxes will tend to turn to chipmunks or something else. Your chance of being eaten by a fox, then, if you are a rabbit, is greater if rabbits are common than if rabbits are scarce. This serves as a sort of automatic brake on an excessive increase in rabbit numbers.

Density relations are particularly important in the case of parasitism and disease. The parasite always has the problem of getting from host to host—and the more scattered the hosts, the greater the problem. The complicated life histories of most parasites and the frequent alternation of hosts are most easily understood in these terms. The parasite may have a nice, comfortable life in the intestine or the blood or the liver of some host, but this very specialized life requires very special adaptations which make it difficult or impossible for the parasite to live anywhere else. How, then, do you get from one host to another, especially if your host species is widely scattered, with individuals rarely coming in contact with each other?

Parasites have solved this problem in a wide variety of ways: through having resistant spore stages that can wait until eaten by some possible host; active stages that can leave one host and move about in search of another; and often through adaptations to a series of very different hosts. Mosquitoes, for instance, serve as vectors or alternate hosts for a variety of human parasites, including many viruses, the protozoan *Plasmodium* that causes malaria, and the filarial worms that cause elephantiasis.

The mosquito provides a means of getting from one blood stream to another, and from one host to another, possibly miles away. But the ability to live, at one time in human blood and at another in the body of a mosquito, requires two very different sets of adaptations on the part of the parasite—the sort of adaptations that can most easily be understood in terms of long evolutionary histories.

One can see that the more numerous and more crowded the hosts, the simpler the transmission problem for the parasite, hence the greater the likelihood of infection for any particular host. As hosts become more abundant, then, the likelihood of severe infestation by any particular species of parasite becomes greater, and the damage becomes greater until it is more than the host species can take—so the host starts to decline in numbers until it becomes scattered again.

The biological community is an infinitely complex network of relationships like this which act and interact to keep the numbers of different things within a fairly stable balance. But man, in many ways, has upset these interacting systems. In the first place, his own numbers have got out of balance. The growth of human numbers was probably a very slow process for a very long time: pre-humans and humans gradually becoming more numerous as they gained social and technical skills, as they learned to work together and to make better tools. But man, in the hunting or food-gathering stage, was still very intimately a part of the biological community in which he lived.

This changed with the development of agriculture, with the Neolithic revolution. When man started to plant his vegetable food and herd his animal food, he entered into a new sort of relationship with the biological community—he began to shorten and modify food chains, to alter and simplify the interacting balancing systems. In clearing forest and planting crops, killing predators and herding prey, he was restructuring the community at the same time that he was vastly increasing his own food resources.

This meant that men could increase greatly in numbers and

could aggregate into villages where they could continue to live in the same place for long periods of time. Among other things, this altered man's relation to his parasites. I suspect that the human contagions, the diseases that are "catching," that can pass directly from host to host without vectors or intermediate stages, started to appear during this period. It is hard to imagine diseases like measles, smallpox, gonorrhea, syphilis, tuberculosis, being maintained in the Old Stone Age. There were not enough hosts, and the hosts would not come in contact often enough, to keep the pathogens going. Man got an increased load of disease along with his increased food supply. Even the non-contagious diseases, like malaria and yellow fever, could take on a new importance.

The contagious diseases today cannot maintain themselves in scattered hunting or food-gathering populations. They are diseases of civilization, often introduced into hunting populations by outside contact, where they may cause fearful epidemics before they disappear again, but they do disappear unless the outside contact is frequent enough. These diseases have a long history in the Mediterranean and in the thickly-settled parts of Asia—they started in the very regions where the Neolithic revolution started—and they have spread to the rest of the world only in modern times.

The disease picture of modern man, then, can be understood partly in terms of his unusual population density, partly in terms of his alteration of community relations. But man's global movements must also be taken into account. A shift in the behavior of a pathogen, resulting in a "new disease," can occur in any part of the world, and the disease can be rapidly spread to all parts of the world. The influenza epidemic of 1918 is an example. The contagions, particularly, are mostly cosmopolitan. The pathogens with indirect transmission are more apt to be restricted, tied to a particular biological community by their complex host relations. African sleeping sickness, for instance, cannot spread to places where the tse-tse fly vector does not exist. And the distribution of malaria is limited by the distribution of suitable species of *Anopheles* mosquitoes—with the difference that *Anopheles* mosquitoes are much more widely spread around the world than tse-tse flies.

The importance of man's global traffic can hardly be overestimated in trying to understand man's present relations with the biosphere. This is true with regard to man's own diseases, the diseases and pests of his crop plants and domestic animals, and even of diseases in the wild. Usually parasites and hosts, in the course of the slow development of the relationship in a particular community, have achieved a working balance. When the parasite gets introduced into a new community, it may not be able to survive under the new conditions—or it may go berserk. Chestnut trees and chestnut blight get along all right together in the forests of China. But when the blight (a fungus) was accidentally introduced into North America, the result was the almost complete extermination of the American chestnut tree, which had no immunity from the parasite. The chestnuts marketed in the United States now come from resistant tree varieties imported from Asia.

Man has altered the biosphere by his transportation of pathogens. But also, I suspect, man's history has in part been determined by the geography of these pathogens. When European man arrived in America, he brought with him a whole collection of pathogens that he had more or less learned to live with, but that were new to America and that wrought havoc in the American population. One can plausibly argue that smallpox conquered Mexico, not Cortez.

The civilizations of Asia and Europe had had close enough and long enough contact to share most diseases, so that the Asiatic civilizations did not collapse so easily. Africa, in turn, had a whole collection of local diseases that had presumably evolved with man in Africa and that had remained tied to African conditions—yellow fever, many strains of malaria, sleeping sickness, and the like—and these diseases were disastrous for the susceptible Europeans. Tropical Africa was thus for a long time protected by its diseases from European invasion and settlement.

Man, and the plants and animals closely associated with man, thus present an unusual situation for the development of parasitism and disease. But the development of medicine and public health has more than compensated for the threats latent in these circumstances. We have learned how to interrupt transmission through sanitation practices, to induce immunity through vaccination, to kill parasites or vectors with special chemicals. In the parts of the world with a highly developed economy, the dangerous infections have come to be very largely under control: ecological hazards to the life of the individual have been drastically reduced.

Death in man, then, is more and more a consequence of internal factors, of the breakdown of the organization of the individual—physiological death. This is a distinctly "unnatural" state of affairs, but I can't imagine anyone deploring it. It does mean, however, that the usual ecological controls on population growth are gone, and the human species has started on a dizzy spree of multiplication that, if not checked in some way, may in the long run be disastrous both for man and the rest of the living world. We have embarked on the enterprise of suspending "natural" controls on one side of the population equation; we must, it seems to me, give corresponding attention to the other, the birth, side.

Parasitism—and disease—can thus be looked at as one kind of interaction between populations in the biological community. We can describe such interactions in terms like food-chains and discuss them in relation to community structure. But we can also take a quite different point of view. Food chains depend on the food habits of the animals that form the different links, on animal behavior. Behavior can be looked at as the cement that holds the community together. Curiously, by an accident of the organization of science, community structure is studied by people who call themselves ecologists and behavior by people who call themselves comparative psychologists, physiologists or, most recently, ethologists. These labels represent different points of view—but the organisms remain the same.

In the next chapter I will look at some of the problems involved in the study of animal behavior. Plants respond to the environment in which they occur; they get food, they reproduce, they grow, they have various methods of defense (think of nettles or poison ivy), they have all sorts of ways of getting about in the world. But instead of trying to make a balanced survey of the behavior of plants, animals and microbes—a very difficult task—let's look one-sidedly at certain aspects of animal behavior.

12. Animal Behavior

It seemed that animals always behave in a manner showing the rightness of the philosophy entertained by the man who observes them . . . Throughout the reign of Queen Victoria all apes were virtuous monogamists, but during the dissolute twenties their morals underwent a disastrous deterioration.

—BERTRAND RUSSELL,
in *My Philosophical Development*

I have spent many hours of my life trying to understand the world in which mosquitoes live by trying to look at the world from the mosquito point of view. It is not easy and I can't report much progress. It is a world that I can know only indirectly through inferences from the way mosquitoes act; from experimentally altering the environment to see what happens; and through trying to measure things I can't perceive directly with various sorts of instruments.

I go on the assumption that the mosquito's world is made up of the same general sort of elements that my world is. If I study the sense organs of a mosquito, I can presume from the way they are made that they serve for perceiving light, odors, tastes, and vibrations, though I can't always be really sure of the functions of the sense organs that I find. Their sensitivity is clearly quite different from the sensitivity of my sense organs, and when I try to make mosquitoes carry out the normal process of living—bite, mate, lay eggs—under laboratory circumstances, I frequently end up completely frustrated. I've overlooked some factor that is essential before the mosquito will carry out that particular activity.

Sometimes I've thought about abandoning mosquitoes, taking up instead the study of mice or monkeys. With a fellow mammal I might have some chance; we would have, essentially, the same kind of eyes, ears, noses, touch perception. I might have a chance at understanding the world of a mouse. But I never have made any serious study of the behavior of mammals. It wouldn't be any easier than the study of insects, I am sure; though I suspect the difficulties would be of a different kind. I think I could do a better job of reconstructing the perceptual world of a mouse, but there would always be the danger of interpreting the mouse's behavior in terms of my own feelings and reactions—the danger of anthropomorphizing: attributing human emotions and feelings to other animals. There is no danger whatever of anthropomorphizing when you are studying mosquitoes—human reactions just won't fit at all in the insect world. Since in many cases you can't even figure out what the mosquito is reacting to, there is little likelihood of sympathizing with the reaction.

What makes a mosquito bite, for instance? Or perhaps I should say, how does a mosquito locate a suitable host? Quite a few scientists have fussed with this question for quite a few years, but the answer still is far from clear. One would first suspect smell. Insects in general have an extraordinarily keen sense of smell; the paired, plumed antennae are primarily organs of smell because apparently insects live in a sort of three-dimensional world of odors hard for us to imagine. Some male moths can locate a female of their own species a mile or so away and fly directly toward her. Butterflies locate the proper species of plant on which to lay their eggs by smell. Insect behavior, over and over again, appears to be governed by odor gradients in the environment.

But efforts to explain the behavior of a biting mosquito by smell have generally failed. Sweat, blood, feces, all efforts to abstract the smell of an animal from the living animal itself have failed to arouse mosquito interest, or have aroused reactions no different from those aroused by a slice of apple or a wad of cotton moistened with perfume.

Several different people who have worked with the problem have found that a hungry female mosquito will react to warmth or moisture or both. A Dutch scientist discovered some years ago that mosquitoes would persistently try to bite a glass tube covered with wet filter paper and kept warm by circulating warm water. But it hardly seems possible that this attraction could operate at any distance. A bit of "dry ice" (frozen carbon dioxide) introduced into the cage would increase the frenzy of the mosquitoes trying to bite the "artificial arm." Many sorts of experiments show that hungry mosquitoes react positively to an increase in the carbon dioxide in the air (every animal constantly breathes out a higher concentration of carbon dioxide than it breathes in) but again it is difficult to imagine this attraction operating at any distance from the animal.

During the Second World War, I was involved in a great deal of work on mosquito repellents. Certain plastic solvents were discovered to be quite effective in inhibiting mosquitoes from biting, and much of the work was a simple sort of experimentation aimed at trying to find out which of these were most effective and for how long. I have no idea how they worked—why they stopped the mosquitoes from biting—and I don't think anyone does. One of the most effective of these substances is di-methyl-phthalate. I remember once working in a flooded tropical river valley where mosquitoes were incredibly abundant. I was stripped to the waist, but I had smeared myself with di-methyl-phthalate. Swarms of mosquitoes would hover around me, coming within an inch of my skin, but never lighting, never trying to bite. What had attracted the mosquitoes to me—and what stopped them an inch away?

Here is a simple behavior problem which I have outlined in some detail just because it is simple, and yet exasperatingly difficult. I am sure this aspect of mosquito behavior can be "explained." There are endless ideas to be tested, and I have probably overemphasized our present ignorance. It isn't one problem, it is a series of problems: the problem of the factors, physiological and environmental, that put a mosquito into a state where it is ready to bite; the problem of factors in host location that operate at a distance; and the problem of factors operating in the immediate vicinity of the host. And every behavioral problem seems to be similarly complex, composed of a series of subproblems that have to be teased out and analyzed separately.

I suppose behavior can always be looked at as a pattern of stimulus and response—sometimes in terms of a simple reflex like pulling back the hand as a response to the stimulus of a burn; sometimes extremely complex like writing a book in response to goodness knows what collection of stimuli. To gain any understanding of the behavior, we have to know something about the stimuli, which leads up again directly to the problem of the environment and to the ways in which parts of the environment are perceived. And this leads in turn to the problem of the nature of the real world, the nature of reality.

Man has been preoccupied with this question of the nature of reality for so long that he has succeeded in getting the issues thoroughly confused. One way out is through the philosophical position of Plato—that the real world is the world of the mind, of ideas. But this, in science, doesn't help us a bit. It is a basic assumption of science that there is an external reality, and that this external world can be described, analyzed, "understood."

This external world is the environment in which organisms live—or is it? I suppose we might define the environment in three different ways: as including only the elements perceived by the organism; as including all elements which affect the organism, whether perceived or not; or as including all elements that can be detected or inferred, whether they influence the organism in any way or not. We might call the first the perceptual environment of an organism; the second, the effective environment; while the third, I suppose, is the total reality that worries the philosophical mind.

Biologists are mostly concerned with the first two types of

environment, leaving the third to the physical scientists—and the philosophers. By setting up a radio receiver in a forest (or anywhere else), we can translate a certain kind of radiation into sound, which we can hear. The radiation is always there—nowadays much of it music or speeches which human transmission stations have contributed to reality; but always there is static, radiation from the stars and from electrical disturbances in the atmosphere. But we know of no way that this affects any organism; of no way in which an organism can perceive it except with the intervention of instruments devised by man. The biologist never thinks of it as part of the forest, even though it is there.

For our present purposes, then, we can forget about total reality. In ecology we are concerned with the effective environment—the parts of this total reality that in some way act on the organism. On the other hand, in analyzing behavior, we must study the perceptual environment—the factors in the external world that the organism perceives and reacts to. These two kinds of environment are probably similar for most animals: they need to be if the animal is to get along in the world. If man had evolved in an environment in which electric wires carrying high voltage currents were common, he would presumably have developed sense organs enabling him to recognize "live" wires. But this isn't necessarily the only solution. Pathogenic microbes are an important part of man's effective environment, but he can't see them, smell them or hear them. He has, however, developed resistance to their effects. Maybe, in the case of the previous example, man would have developed immunity to electric shock.

This sounds like quibbling, but I think it is important because we so easily make the mistake of assuming that our world —the world perceived by human senses—is the "real" world. A particular bit of forest is a very different place to a caterpillar, a bird, or a man living there. We naturally describe the forest in the way we see it—which works for most human purposes. But it doesn't necessarily work if we are trying to understand the behavior of the bird or the caterpillar.

Students of behavior make much of the sin of anthropomorphism, but I have never been able to get as upset about this as some of my colleagues. In fact, I think the effort to avoid the use of human terms in describing animal behavior often produces not clarity and objectivity, but inhuman and unreadable prose. When we speak of "angry bees" we don't mean that individual bees have the sensations of individual men in an angry mob. We are using a useful metaphor—and if the behavioral scientists removed all metaphors from language, there would not be much left.

The sin of confusing the human perceptual world with all of reality, or with the perceptual or effective environments of other organisms, seems to me much more dangerous than anthropomorphism in the usual sense—and much more difficult to avoid. We have no way of finding out about total reality, or about the environment of any other organism, except through our sense organs and through instruments that extend our senses (like the microscope) or that translate things we can't perceive into things we can (as a Geiger counter translates cosmic radiation into clicks or into markings on a dial). We have devised some marvelous instruments and discovered many extraordinary things. But there may still be aspects of reality to be discovered, and we know that the perceptual world of other organisms includes elements to which we are entirely blind. Sometimes it is fairly easy to keep this in mind, particularly when we are studying organisms in an environment that is foreign to us—in the sea, for instance. But sometimes it is very hard to remember that the world built by our senses may be very different from the world of some other animal that we see in the forest, or in our own backyard.

Most students of behavior group sense perceptions into four classes: visual, auditory, chemical and tactile. We live in a primarily visual world, and so do a great many other animals, but it is clear that there are enormous differences in the way things are seen by different kinds of animals. Apparently man and some of the other primates are the only mammals that can distinguish color as such—the color of a red flag is meaningless as far as a bull is concerned. Day-flying birds, fish, insects, squids and octopuses can, like man, distinguish colors—at least some species, that have been carefully tested, show an ability to discriminate between some colors and shades of gray of corresponding intensity. Whether they "see" colors the way we do is another question. If we were color blind, and other animals not, how on earth could we imagine what it was that they were seeing and reacting to?

From many experiments, it is quite clear that different animals perceive different parts of the light spectrum. Light is measured in terms of ångstrom units—one unit representing a wave length of a hundred-millionth of a millimeter. We perceive light with wave lengths between about 4000 Å and 7200 Å; in color terms, from violet to red. Light with a wave length of more than 7200 A we call infrared; of less than 4000 A, we call ultraviolet.

A honeybee perceives light from about 3000 Å to 6500 Å, which means that it can distinguish colors in the ultraviolet that are invisible to us, but that it is blind to reds that we can see. Insects in general, insofar as they have been tested, see farther into the ultraviolet than we do, which has many consequences. We can make photographic plates that are sensitive to ultraviolet light, and from this it turns out that many things, like flowers and butterflies that seem plain white to us, have distinctive patterns invisible to us but significant to insects. The whole business that I have called ecological coloration—protective colors, mimicry, signal colors—needs to be looked at in terms of perception. What we see is not necessarily what other animals see.

Animals also differ enormously in their ability to see things at a distance, or to see things in faint light. Apparently a dime on the sidewalk below would be perfectly clear to a hawk cruising at the height of the Empire State Building, and an owl can find his way around in light so faint that we are quite blind.

Light, we have gradually learned, is only one aspect of a vast range of phenomena that we can lump together as radiation, the transference of energy across empty space at a constant speed—the speed of light. All of these forms of radiation have many things in common, including a measurable wave length. As wave lengths become longer, we move from visible light to infrared, to a little-known kind of radiation called "rest rays," to heat waves, and finally to the long waves of radio. In the other direction we move from ultraviolet toward x-rays and finally gamma rays and cosmic rays. Some of these radiations that we cannot detect directly with our unaided senses have important biological effects: a whole field of knowledge is growing up under the label "radiation biology."

I wonder, sometimes, whether other animals can perceive parts of this radiation spectrum that we can't. There are great differences in the perception of animals in the immediate vicinity of the light spectrum, but what about much longer or shorter rays? One of the few instances I can think of where we have begun to get a glimmering of this is in the case of the snakes known as pit vipers—rattlesnakes and the like. The "pit" on the head of these snakes is an organ for detecting heat radiation—and it seems that the striking of the snake is guided by radiated heat rather than by sight.

When we move from radiation to sound—to mechanical vibrations in air, water or other substances—the perceptual world of many other animals is utterly different from our own. The most striking and carefully worked out case is that of bats navigating by means of echo-location. A bat in flight is constantly emitting a series of very high frequency sounds and hearing the echo that bounces back from solid objects, so that it can find its way swiftly and accurately in complete darkness.

Individual humans vary somewhat in their sensitivity to sound, but the average range of hearing is from about 20 vibrations per second to 20,000. The bat in echo-location is using sound waves with a frequency of around 100,000 vibrations per second. Dogs can detect a much higher pitch of sound than men can—as is clear enough from the supersonic dog whistles that can be used for calling your pet.

The differences in the perceptual worlds of sound were brought home to me forcefully by the story of a friend who had been studying rabies infections in cattle in southern Mexico. Bats have become implicated in rabies in several parts of the world, and vampire bats (which live by lapping the blood of other mammals) would be logical vectors, and apparently were the vectors in this case. But my friend wondered whether dogs might also be implicated. He was lying awake in his hammock in a country rancho one night when a vampire swooped in, and he saw a dog that had been sleeping soundly beneath his hammock wake and dash out, tail between legs.

"Of course," he suddenly realized, "to a dog, the silently approaching vampire bat would sound like a boiler factory falling in out of the sky." He ruled out dogs in relation to bat rabies.

It's a nice story, but it may not be true. Dogs, certainly, can hear higher frequencies than men, but I don't think their hearing extends into the region used by bats for echo-location.

With many animals, it is difficult to tell whether they perceive a sound by "hearing" it or "feeling" it. Many insects, for instance, respond to sounds, even though they seem to have no special sound-receiving organs. Some insects, however, have a special sound-receiving organ, a tympanum, sometimes located on the sides of the body, sometimes on the legs, and with such insects sound is clearly important, especially for sexual recognition, as with crickets.

We have only recently come to appreciate the importance of sound in water. Sound travels faster and more easily in water than in air and is an important factor in the behavior of fish—as well as in the behavior of such marine mammals as dolphins and whales. Slight changes in water pressure and slight water disturbances are more significant for marine animals than similar changes in the less dense medium of air are for land organisms. The lateral-line organ of a fish is an elaborate sense receptor for slight pressure changes, which seems to have no counterpart among land animals. Hearing and touch blend in a sense organ of that sort, and I suspect it would be most logical for animals as a whole to combine sound and touch perception under the general heading of perception of mechanical disturbances in the animal's surroundings. This would give a threefold division of the perceptual environment: radiation, mechanical disturbance, and chemical composition (which we detect through taste and smell).

We can taste and smell, but clearly we live in a poor and limited chemical world compared with many animals. We can easily see the difference, in this respect, between our world and that of our dogs—all of the things that have to be investigated and savored in the course of a walk; the canine ecstasies borne by the breezes during a drive through the country.

We have, it seems to me, made little progress in our studies of the world of smell, perhaps because it is so unimportant in our own perceptual environment. We haven't as good a way of classifying smells as we have for classifying colors. We haven't any good way for measuring the intensity of smells, for studying the possible consequences of mixing smells. Where our own senses fail or are unreliable, we have frequently been able to develop sensitive instruments for measuring things, but we haven't made much progress in instrumentation for the study of odors.

Smell is particularly important in the insect world. The elaborate, paired antennae are primarily organs for detecting extremely faint odors and for determining the direction of their source. The silkmoths are always mentioned in this connection. Anyone who has collected cocoons in the winter and let them hatch in the spring has noticed that when a female hatches, males soon appear outside a screened window, bumping against it in their efforts to get in. A male cecropia or a male luna can apparently detect a female of his own species at a distance of a half a mile or more and, following up the smell gradient, locate her.

Sexual behavior in insects turns very largely on smell, and

so does food behavior. And the feats of insects in the two cases are equally incredible from our point of view. The caterpillars of butterflies often have very restricted food habits, living on only one species, or a few closely related species of plants. The butterfly, in finding the right species of plant on which to lay its eggs, relies on smell. Each of the thousands of kinds of herbs, shrubs and trees must have its characteristic odor which serves as a cue for the insects that feed on it. The forest to us presents an endlessly varied series of patterns of sights—of leaves and flowers and stems, of shapes, movements and colors. With appropriate sense receptors, it must present equally rich and diverse patterns of odors. Each kind of plant could be quickly identified, not by the form of its leaves and flowers, but by the smell of the essential oils, alkaloids and other chemical constituents. How do we study this aspect of the world? It is as though we were all blind and trying to understand the significance of light.

No matter where an insect learns to live a hidden life, some other insect learns to find it. The parasitic wasps show endless skill in this. Some species are hyperparasites, living on parasites that are already in some other insect. The ovipositing wasp can always tell which hosts have inside them the parasites on which her larvae can live. There are wasps that specialize in parasitizing wood-boring larvae, locating the hosts deep in tree trunks, presumably through smell, and boring in with their long, drill-like ovipositor to lay the fatal eggs.

With the visual world, we can see the variety of adaptations —signal colors to attract appropriate mates or pollinating insects, protective colors and patterns to hide from enemies, warning colors. We can get some glimmering of signal odors and warning odors, but I have often wondered if there aren't all sorts of protective and concealing odors, a host of adaptations quite unknown to us.

Smell in the sea is clearly as important as in the forest, though here smell and taste merge. Is it some chemical gradient that leads the eels from Europe and America to the Sargasso Sea to reproduce, and that leads the larval eels back to the appropriate continent? Is it the taste of the water that guides the salmon to its home river, that tells it which branch to choose until finally it ends in the mountain stream where it was spawned years before?

The study of behavior is thus inextricably bound up with the study of perception. If we go back to our beginning and look at behavior in terms of stimulus and response, we are concerned with the stimulus part of the sequence in studying perception. This is about one-third of the story because, to get a description of a particular behavioral pattern we need to analyze not only the stimulus that starts it and the response with which it is completed, but also the mechanism that comes between the receiving and the effecting. If we look at an organism anatomically, the sense organs in general are the receivers, the muscles the effectors. But what happens in between?

This turns out to be a very complicated business. In most animals it involves both a nerve system and an endocrine, or chemical, system. Perhaps we can skip over these mediating mechanisms here, because they are clearly physiological, involving the study of the insides of the animal. This shows the absurdity of my attempt to divide "skin-in" and "skin-out" biology. We can hardly discuss the behavior of an animal intelligently without taking into consideration the structure and functioning of its nervous system, or without knowing something about hormones, the chemical "messengers" that control so many activity states. But we have to divide up knowledge somehow, and for my present purposes I'll stick to the skin as the most convenient dividing line in biology.

If we move to the third aspect of behavior, response or action, we at once come across the problem of instinct versus learning. I suspect that this is a vocabulary problem more than anything else, but this doesn't make it any less serious and it has become the focus of a full scale war of words. It all started, I suspect, with the tendency of some students to oversimplify human behavior—to talk about human actions in terms of aggressive instincts, herd instincts, or things like Thorstein Veblen's instinct of workmanship. The result is that instinct has become a nasty word with many psychologists; it is like a four-letter obscenity, and they go to all sorts of trouble to avoid it. Certainly it is very difficult, and apparently almost always misleading, to discuss human behavior in terms of instincts. But it is equally difficult to discuss insect behavior and not use the word.

Actually, instinct and learning form a false antithesis. It is the old problem of nature and nurture, of heredity and environment, in a new form. It is another of the interminable illustrations of the fallacy of either-or. Perhaps one could say that every behavioral pattern has an innate, structural basis, which always finds expression in some environmental context—and that the degree to which the environment may influence behavior, or the degree to which previous experience (learning) may influence it, varies greatly. It varies depending on the kind of animal, and on the kind of behavior. One can then talk about the degree of modifiability of behavior, and about the ways in which it becomes modified.

With invertebrates in general, behavior is pretty thoroughly stereotyped—instinctive. Some degree of learning can be demonstrated with all sorts of creatures, particularly in the laboratory. It might be better to call it conditioning, rather than learning, because for the most part it involves shifting a response through simple reward or punishment. But most invertebrates live out their lives with a set of behavioral responses as fixed and unvarying as their anatomy. The behavior, like the anatomy, is characteristic of each species, and it is inherited in the same way—that is, it is under genetic control.

It is with the vertebrates that the instinct-learning argument becomes important; the major groups of living vertebrates represent a sort of evolutionary sequence in this respect, with learning becoming more important as one moves from fish to amphibians to reptiles to mammals—with the birds in this respect, as in others, not fitting neatly into the series. There is a corresponding sequence in the relative bulk—relative importance—of the brain in these vertebrate groups.

A similar sequence in the importance of learning can be made within the mammals. I have heard Frank Beach illustrate this with the example of copulatory behavior. Mice, on first reaching sexual maturity, will copulate "instinctively" (my word, not Beach's). That is, they will go through the sexual act for the first time in about the same way as at subsequent times. With dogs there is a considerable element of learning—the first copulation will be awkward, performance improving with experience. With monkeys the learning element is still greater, and if young monkeys are raised in isolation from old, experienced individuals, there may be a considerable period before proper copulatory behavioral patterns are learned.

In the Yerkes Laboratory at Orange Park, Florida, young chimpanzees were raised in isolation from older, experienced individuals. When such inexperienced individuals of opposite sexes were put together, all sorts of sexual experimentation followed—but it seemed to be sheer chance if an individual learned proper copulatory behavior, though they learned quickly enough from experienced individuals. In other words, in these higher primates, the details of copulatory behavior depended entirely on learning. One can argue that there was still a general sexual instinct—contemporary psychologists would probably call it a "drive"—but the actual pattern of primary sexual behavior was entirely controlled by learning.

Learning is of such overwhelming importance in human behavior that one can sympathize with the psychologists who would like to forget about instinct entirely, or restrict it to the first sucking and grasping reactions of newborn infants. Students of animal behavior, on the other hand, tend to look for vestiges in man of the innate patterns that they find everywhere—and I suspect that these vestiges of instinct are still with us, however deeply buried under the cultural overlay.

When we discuss instinct, conditioning, learning and the like, we are concerned with the mechanisms of behavior—with the kinds of structures from which the actions of an animal are built. The same sort of process, learning or instinct, may serve many different purposes. If we examine the behavior of an animal in terms of its purpose, its goal, its function, we get involved with a quite different system of analysis. We then are interested in trying to find out what sort of behavior serves the animal for reproduction, for food-getting, for the disposal of waste materials, for dispersal, for defense or protection.

But these different goals, which seem so distinct when we work out a logical scheme on paper, often seem all mixed up when we start watching what an animal actually does. The animal has to have food, has to have a mate, has to be able to get around in the world, but these different ends may all be served by the same general kinds of actions. It is difficult to cut up the process of living into neat and separate pieces that can be laid out on a table and compared or described. We can cut up animal carcasses easily enough and describe the parts—bones, guts and muscles. But the living, acting, ever-changing organism is more elusive. That is why, essentially, we can talk about anatomy so much more easily and positively than we can talk about behavior.

When we look at ourselves, at man, and wonder about the biological basis of human nature, we quickly become involved with the question of the origins of our social interactions. Man, obviously enough, is a social species. When we look at other animals, we see all sorts of evidence of similar social behavior —flocks of birds, schools of fish, colonies of ants. Yet we also see animals that live most of their lives quite separately from others of their own kind. How does man's social behavior compare or contrast with that of these other animals?

When we try to fit social behavior into a logical scheme, we find a number of difficulties. It might be either instinctive or learned or both—its basis, in fact, may be quite different in different animals. And, obviously, it may serve many different purposes—food-getting, defense, reproduction. But instead of worrying about the classification, let's look at the animals.

13. Social Life Among the Animals

Gazing out at me with myriad eyes from their joyless factories, might [the bees] not learn at last—could I not finally teach them—a wiser and more generous-hearted way to improve the shining hour?

—LOGAN PEARSALL SMITH, in *All Trivia*

In a very broad sense, any sort of continuing interaction among individuals of a particular species could be called social behavior. In this sense, social behavior is widespread in the animal kingdom. There are many kinds of colonial animals, like the corals, in which individuals aggregate to form a large and continuing mass; there are many instances of continuing association between individuals of the two sexes of a species; and there are many kinds of parent-offspring relations.

But we generally think of social behavior as something more complicated, more formal, than mere aggregation; something involving specialization and cooperation, social structure of some sort. When we look for something comparable with our complex human societies, we think first of the social insects, of the termites, ants, bees and the like. Here, plainly enough, large numbers of individuals are carrying out all sorts of diversified tasks, working closely together for the good of the group. And the more we study these insects, the more similarities we find with our societies. We find caste systems and work specialization, agriculture, food storage, slave-making, warfare, complicated construction. We find ants "milking" their aphids and taking care of their "pets." We find a wonderful world that seems tantalizingly familiar—though it is utterly strange.

There are many interesting books on the social insects, and I don't want to get into the details of their lives here beyond what may be necessary to underline their difference. In the last chapter I emphasized the difference between the perceptual world of any insect and that of any vertebrate; and the difference in the way their behavior is organized, which turns on the fact that their nerve systems are quite differently arranged. The whole insect world is a topsy-turvy place—as you might expect when you look at the way insects are built. They wear their skeleton on the outside, like a suit of armor. For breathing, air is carried directly to every cell through a system of tubes which start from a series of portholes along the sides of the body. The blood, instead of being confined in vessels, is loose in the body cavity and kept in circulation by an open-ended pump, a "heart," in the middle of the back. The central nerve cord, on the other hand, goes down the middle of the belly. With a body structure so different from that of a vertebrate, why shouldn't the behavioral system be built on quite different plans? I think it is. Consequently, when we see similar behavior in insect and vertebrate, we are dealing with an analogy in which the similarity is quite superficial—though nonetheless instructive.

The parallels between insect societies and human societies are simply another example of the fact that similar-looking things in nature often have quite different origins. In this case, the two things are so completely different that I sometimes think it is unfortunate that we call them both social. All insect societies are essentially gigantic single families—the ants in an ant hill are all sisters and brothers. All human societies are groupings of many families. And this alone seems to me to make all the difference in the world.

William Morton Wheeler has calculated that social organization in the insects has started, quite independently, at least 22 different times, in 22 different groups of insects. In each case the basis is a continuing association between parent and offspring: the young staying with the female parent, or with both parents, to form a cooperating group. In the termites, ants, wasps and bees this simple idea of parent-young association has developed tremendous complexities.

Ants, wasps and bees are all rather closely related—they belong to the insect order Hymenoptera. Social organization in each has an independent evolutionary history but with many similarities. The termites are something else again. They are relatively "primitive" insects, distant cousins of cockroaches. It is as though, with mammals, very complex societies had been developed by both the marsupials (opossums and kangaroos) and the primates. Understandably, then, the two kinds of societies show many basic differences.

Termites are often called "white ants"—it is said because they are not ants and not white. There are several thousand kinds of termites, mostly in the tropics. Only 38 species are known from the United States and only one or two of these get very far north. They are notorious, however, because they may undermine foundations of houses, eating out the timber, so that termite extermination and prevention of termite damage has become a considerable business.

Termites have incomplete metamorphosis. That is, as with cockroaches, grasshoppers and true bugs, the insect that hatches from the egg is similar to the adult, changing mostly in size and proportions with the successive molts in growing up. Ants, bees and wasps have complete metamorphosis; the insect that hatches from the egg is a larva, a grub, utterly different from the adult, and the adult form is assumed only after a final resting stage, the pupa.

This difference has many consequences in social structure. The larvae of ants and other social Hymenoptera are helpless and must be protected and fed until they have grown up.

Young termites are active from birth. In fact it appears that worker termites in general are perpetual adolescents, prevented from ever really growing up to reach sexual maturity by a sort of social hormone that circulates through the society by means of the continual food exchange that takes place among the individuals of a termite colony. It has been aptly said that the termite social organization is based on child labor.

Termites live on dead wood, on cellulose, extremely difficult stuff for any animal to digest. The termites have solved this problem, as I mentioned earlier, by a commensal arrangement with protozoans that live in the gut: the protozoans apparently have evolved with the termites, and each species of termite has its species of protozoan, the two being completely dependent on each other. But the wood, even so, is only partially digested by its passage through the gut of a single termite, and the feces are eaten over and over again. In addition to this kind of food exchange, the termites produce waxy secretions which serve as food for fellow termites, and the members of a termite colony are constantly licking each other, providing a close bond of food relations that keeps the whole colony together. Wheeler coined the elegant word *trophallaxis* to describe this social food exchange.

Each termite colony has a "king" and "queen"—better called "primary reproductives" since their function is to produce eggs rather than to rule the colony. Both sexes are also represented among the workers and soldiers. If either the male or female primary reproductive is removed from the colony, a few workers of the appropriate sex will start to continue development to reach sexual maturity. Now the primary reproductives are constantly being licked by the workers around them, and from various sorts of experiments it appears that the secretions of these individuals contain substances which inhibit other individuals of the colony from becoming sexually mature.

Furthermore, termite colonies generally contain a special caste called soldiers, individuals with big jaws or other special adaptations for the defense of the colony. The proper balance of numbers between workers and soldiers is apparently also maintained by a social hormone system. If all the soldiers are removed from a small colony, more of the young will promptly start developing into soldiers. The only likely explanation is that the number of soldiers being developed is controlled by the amount of soldier-inhibiting substance released into the food exchange system of the colony by the soldiers actually present. The whole system is analagous to the hormone system of a single animal through which the relative growth of different parts of the body is promoted or inhibited by chemicals, by hormones, produced by the ductless glands like the thyroid or the testes.

It has often been argued that the colonies of social insects should be regarded, not as aggregations of individual insects, but as "superorganisms." The individual termite or ant, sexually sterile, has no biological meaning apart from the colony to which it belongs. Individuals of these insects, as a matter of fact, cannot even be kept alive for any length of time if they are by themselves—which doesn't make studying them any easier. The significant biological unit, from the point of view of evolution, of reproduction, of ecological relations, is the colony as a whole. The individual termite has a biological meaning comparable to that of the individual cell in the vertebrate body. The whole system of social control through trophallaxis gives support to this point of view.

When we look at the social Hymenoptera, at the ants, bees and wasps, we find equally tight colony organization. The honeybee from many points of view is the most thoroughly studied of any insect because of its importance in human economy and because of the sheer fascination of trying to understand its complicated behavior.

The beehive includes a queen—again better called a primary reproductive—many thousands of workers (all sterile females) and a few drones or males. The workers all look alike—there is no special soldier caste. The drones have a quite different function from the reproductive male termite, since they do not fertilize the queen of their own colony. The queen bee mates only once, on a special mating flight high in the air soon after she has first emerged. She acquires, on this mating flight, a store of sperm which, kept in a special organ called a spermatheca, lasts her for life.

The queen bee, like the primary reproductives of all social insects, lives a life of constant egg-laying—in this case into cells made by the workers. If a particular egg, when laid, is not fertilized by a sperm from the spermatheca in passing down the oviduct, it will develop into a male, a drone. If it is fertilized, it will develop into a female, either a worker or a queen, depending on the kind of food the developing bee grub is fed by the workers.

There are still many things about the coordination and organization of the beehive that we do not understand, but Karl von Frisch, in a series of brilliant studies that have been confirmed by other workers, has provided great insights into the "language of bees." "Scout" bees, returning from the field, communicate to fellow workers the direction, distance, and amount of food they have found. They do this through "dancing." If the food is at any distance, the bee goes through a "waggle dance," running on the comb through a sort of figure-eight turn and stopping in the middle to waggle the abdomen; the number of turns roughly indicates the distance of the food—the fewer turns the farther away. Direction is given by straight runs of the waggling bees over the comb: directly upward if the food is in the same direction as the sun, downward if away from the sun, or at an angle from the vertical corresponding to the angle of the sun. The more vigorously this dance routine is carried out, the larger the amount of food. The kind of food, of course, would be directly indicated by the sample of pollen and nectar brought back. Ants and other social insects must have equally precise ways of communicating information, but the code has not been broken yet.

The insect societies have been in existence for a very long time. Some of the fossil ants from the Oligocene, sixty million years ago, differ hardly at all in appearance from ants living today, and presumably also differ hardly at all in social behavior. The various insect societies unquestionably are biologically successful both in terms of survival and of abundance. One sometimes wonders whether human society, with its very different origins, is on its way towards a similar rigidity, a similar subordination of the individual to the group, a similar sort of superorganism character. Some people argue that human societies already are superorganisms. I don't think so, but maybe this is wishful thinking.

Man is a vertebrate, a mammal, a primate. To understand the development of human behavior, we must look at the behavior of these animals. In particular, we must look at the various kinds of social behavior that we find in these animal groups—in the vertebrates in general, in the mammals more narrowly, and most particularly, in our surviving relatives the primates.

We find a wide variety of social behavior among the vertebrates. The variety is reflected in the list of words that we use as labels: school, herd, pack, flock, as well as such special things as coveys of quail and prides of lions. Vertebrates, then, have frequently discovered the advantages of living together in cooperating groups, and these groups represent a variety of independent evolutionary experiments—mostly, I think, with little relevance to the evolutionary experiment that led to human society. The groupings sometimes are only for one period in the life cycle, sometimes only for part of the year, sometimes involving only one sex.

With the vertebrates—as with most animals, for that matter—we have three major classes of individuals within the species: males, females and young. If we look at social behavior in its broadest sense, as continuing interaction between individuals of a particular species, this means that we have six basically different kinds of interaction: male-female; female-young; young-young; female-female; male-male; and male-young. Looking through the vertebrates, we can find all sorts of interesting examples of extreme development of each of these kinds of interaction, as well as all sorts of combinations.

With many fish and a few birds, the care of the young devolves entirely on the male parent, so that the male-young relationship predominates. With many fish also only the young organize in schools, making the young-young interaction the chief social bond.

All mammals are committed to the female-young interaction, since the young are completely dependent on the mother for food until weaned. All mammals must also, periodically at least, show male-female interaction in sexual contact, since fertilization is internal, and since the female mammal has no organ like the spermatheca of the insect in which sperm can be stored for a lifetime. The young are generally born in litters of several at a time, which makes young-young interaction inevitable. Where the young are born singly, an older offspring is frequently still associated with the mother when another sibling is born, so that young-young reactions may occur in this case too.

All mammals, then, by our definition, necessarily show some form of social behavior. The mother-young grouping is universal and forms the simplest possible form of family. Male-female interactions always occur, and quite frequently particular males stay in association with a female, making a somewhat more complex family association. The next step in complication is for a male to gather a harem of several females and remain with them. I would still regard this, however, as a family group even though such groups may become quite large and complex. The next step, the association of several males with females and young, is relatively rare, but is especially interesting from the point of view of the possible animal origins of human social behavior. Such a group can be called social in a larger-than-family sense and necessarily involves all the combinations of interaction among males, females and young. Groupings of this sort are especially notable among the canines —wolves, dogs and the like—and among the primates.

Before pursuing this further, I would like to look at two general aspects of vertebrate behavior which have social implications—territoriality and peck-order, or "social hierarchy."

In 1920, Eliot Howard published a little book, *Territory in Bird Life,* in which he reported observations he had made on the birds nesting in his garden and orchard. He noted that each bird family had staked out a rather definite territory which was regarded, so to speak, as private property, and defended against intrusion by other birds of the same kind. The frantic singing of the early-arriving males serves chiefly as a warning that this particular territory is already occupied.

Many studies of territoriality have been made since 1920, and a great deal has been learned. There are many kinds of territory—for mating, for nesting, for feeding, or for combinations of these. Territory is sometimes defended by individual males, by individual females, by pairs, or by larger social groups. Birds that nest in large flocks, like flamingoes and gulls, may defend a very small nesting territory from intrusion by others in the flock. Territorial behavior of one sort or another has been found in fish, amphibians, reptiles, birds and mammals—though territorial behavior is not universal in these groups.

Territory is usually defined as an area defended against intrusion by other members of the same species, and is perhaps the commonest cause of intra-specific fighting among animals. Fighting for mates may be commoner, though it is sometimes not clear which kind of combat is involved. The fighting is almost always largely bluff, and someone gives way before real injury is done. The first-comer to a territory usually wins, and the nearer he is to the center of his territory, the braver he is in defense. It is not always easy, however, to determine whether a territory is defended or not. With many animals, individuals or family groups are found staying over long periods of time in particular areas, but there is no direct evidence that these areas are defended. This sort of an area is generally called a "home range" rather than a "territory."

When it is a group—family or tribe—that occupies a particular territory or home range, the shared terrain clearly helps to give cohesiveness, "togetherness," to the group, and thus may be of considerable importance in social behavior. Any mammal is ill at ease outside its familiar range—if I may be anthropomorphic—and tends to stay at home unless pushed away by events inside the range or powerfully attracted by events outside. There is reason for this, since an animal outside its familiar range is much more exposed to its enemies, not knowing where to dodge or hide, and there is some evidence that animals outside their territories, or that are unable to establish territories, are most subject to predation.

I remember watching the process of range-learning in a family of coatis—a relative of the racoon—that I was raising in South America. Coatis are intelligent, gregarious and brave animals: but the babies started out being timid enough, and one could see the gradual growth of self-confidence. At first they would go only a little way from the nest box, scuttling back home at any alarm. But by a slow process of ever widening exploration, they extended their range to cover all the house, finding secondary retreats under sofas and behind books, until presently they were masters of the whole house, secure anywhere.

Territoriality is such a common, easily observed affair that it is curious that naturalists did not become generally aware of it until 1920, when Howard published his book. One can, of course, find a number of observers who anticipated Howard, but their ideas failed to make much impression on students of animal behavior. Similarly with peck-order or social hierarchy. Everyone has noticed that, after a few fights, one dog will give way to another without going through the fight every time; and the cock that rules the roost must have become a cliché a long time ago. But it wasn't until 1922 that this was developed into a formal theory by a Norwegian, Schjelderup-Ebbe, on the basis of his study of the behavior of a flock of chickens. Since then, social dominance has become a fashionable field of study and the subject of any number of doctoral theses and articles in learned journals.

In any flock, there is generally one chicken before whom all others give way, another occupying second place, and so on down to one poor bird that is pecked by everyone. This social order may be stable for long periods of time—and you can see the fascinating experimental possibilities. If you remove the top bird, how long can you keep it away and still have it automatically step back into top position? If the fighting starts all over again, does the original bird still come out on top? What characteristics determine what happens to a newcomer? What about dominance between sexes and among members of the same sex, or among young?

The peck-order is presumed to favor social stability. If there were endless fighting over every bit of food, over every female in heat, over every resting place, the result would be chaos. But order results when every individual knows his (or her) place. Schjelderup-Ebbe was sure he had found a basic principle of nature. "Despotism," he wrote, "is the basic idea of the world, indissolubly bound up with all life and existence. On it rests the meaning of the struggle for existence."

The principle holds quite well for chickens, for children in kindergarten, for thieves both in and out of jail; but its universality has not been confirmed. A dominance hierarchy among individuals has been repeatedly demonstrated with many kinds of animals in captivity, but attempts to describe peck-order among social animals in the wild have not been so successful. This surely is largely because of the immense difficulty of making observations, of recognizing individuals and interpreting individual interaction. But I suspect also that peck-order is in part an artefact produced by confinement—which may be why it shows up so clearly in humans confined in schools or jails. It remains, however, a factor that must be taken into account in all attempts to understand vertebrate social behavior.

Social dominance relations gain particular interest in the

mammals with larger-than-family groupings, in the canines and primates. Unfortunately, we know extremely little about the behavior of the members of the dog family in the wild, for understandable reasons. The best study I know of was made by Adolph Murie of a pack of wolves near Mt. McKinley, which he was able to watch with field glasses for extended intervals. The pack consisted of three adult males and two females, each of these with a litter of five pups. All of the interaction he was able to observe among the adults was obviously friendly, always accompanied by much tail wagging. He once observed a strange wolf attempting to join the pack—unsuccessfully. The stranger was driven away, rather badly wounded; in this case the initiative seemed to have been taken by one of the males of the pack, but this would hardly prove that this male was dominant within the pack.

The most complete study of monkey behavior known to me was made by C. R. Carpenter on the howler monkeys of Barro Colorado Island in Panama. Barro Colorado, a forested hill that became isolated as an island when Gatun Lake was formed in the course of constructing the Panama Canal, has been set aside as a natural history reserve. Hunting is not allowed and many of the animals have consequently lost their extreme shyness. It is a beautiful place to study the behavior of animals of the tropical American rain forest.

The howlers are the largest of the American monkeys (in terms of body weight) and the noisiest. The hyoid bone of the throat is enlarged into a sort of big box, a resonator, and the whole larynx is greatly developed. The monkeys regularly greet the dawn with howling choruses, howl at any intruder into their forests, nowadays howl at airplanes passing overhead. I once learned to make a fair, though feeble, imitation of their sound, and got a deal of pleasure out of howling at them. At this insolence, the big-bearded males would come down low in the trees, grimacing and shaking the branches as they howled back. I had to remind myself how much bigger I was than they were, and that I could undoubtedly run faster.

Dr. Carpenter spent nearly a year on Barro Colorado watching these monkeys every day. He found that there were about 400 howler monkeys on the island, which had an area of a little less than 4000 acres. The monkeys were organized into 23 clans, each with a definite territory whose limits could be recognized by particular trees or by features of terrain. The clans varied in size from 4 to 35 individuals with an average of 17 or 18: 3 adult males, 7 adult females and associated young.

The clans were strictly territorial, staying within a particular area and often following the same preferred pathways through the trees. When two clans came near each other on the borders of their territories, a vigorous howling was set up until one or the other, or both, retreated. There were a few solitary, "bachelor" males. These, when they approached a clan, were shouted off by the clan males. Carpenter watched one case, though, where a bachelor persisted in trying to join a clan over a period of months, and was finally accepted.

Within the clans, Carpenter could find no signs of peck-order, no signs of a single leader. On the move, a male was generally in the lead, but sometimes it would be one male, sometimes another within a particular clan. He found, further, no signs of sexual jealousy. Sex was in the hands of the females; a female in heat would approach one male, and if rebuffed, try another. When a male became satiated, the female went soliciting again. All aspects of howler organization, in short, seemed to be peaceful and communal, within the clan.

One gets the impression that the other New World monkeys, the spider monkeys, squirrel monkeys, Cebus monkeys and the like, show a communal organization not too different from that of the howlers. The Old World monkeys—which belong to a quite different family, with a distinct evolutionary history—show a similar tribal organization, with several adult males as well as several adult females in each group. With these monkeys, at least with the baboons and macaques, there is considerably more evidence of a social hierarchy within the group, with a single, old male dominating everyone else. To understand the behavior of these monkeys, however, we need much more extensive and careful field studies than have yet been made.

If we turn to the animals anatomically most similar to man, the great apes, we find that social behavior turns on family groupings rather than on larger-than-family groups. To be sure, our knowledge of the home life of the great apes is not very extensive. In their native forest the apes are better at avoiding observation by scientists than the scientists are at observing the apes. And I think we have to be very careful about making deductions from the behavior of animals in captivity. It is like trying to describe human nature from observations made in a prison: even the best-run, model prison doesn't illustrate very well the interpersonal relations of men —or men and women.

The best field study of ape behavior again was made by C. R. Carpenter, this time in Thailand, where he was able to study the gibbons in a forest that belonged to a Buddhist monastery where the animals were not hunted. The gibbons, Carpenter found, showed a strictly monogamous family organization: each group consisted of one adult male, one adult female, and one or two young, the young leaving the group with the approach of sexual maturity.

Very little is known about the orang-utan in nature, but it is commonly thought to be the least sociable of the great apes, not even forming family groups beyond the necessary mother-offspring relationship. Orangs that have been brought up in human households, however, seem to be as social as chimps brought up similarly, and it may be that the orang has a more highly developed family life than we have suspected.

Chimpanzees have been studied much in captivity, especially at the Yerkes Laboratory in Florida, and they show many kinds of cooperative, social activity. In the wild, chimps live in groups consisting of from 4 to 14 individuals. But these groups appear to be essentially family groups, composed of a single adult male, several adult females and their associated young. The gorilla in the wild, like the chimpanzee, lives in polygamous family groups, dominated by a single adult male.

This prevalence of basically familial organization among the great apes makes me doubt the direct relevance of studies of their social behavior to the problem of the origin of human social behavior. The fallacy in the idea that man has descended from the apes has been explained so often that there is not much use in laboring it more here. But the idea still lingers, especially in behavioral terms. The apes are, clearly, man's closest living relatives, but this is far from saying that man's prehuman ancestors must have looked and acted like the present great apes. Because John is your cousin doesn't mean that you can learn much about your grandfather by studying John's looks and actions. And the fossil evidence increasingly indicates that the ape line and the human line have followed independent courses of evolution for a very long time.

Man, to be sure, has strong family characteristics, and humans behaving like gorillas, chimps, gibbons or even orangs can be found easily enough. But all human societies that we know involve groupings larger than the family; in particular, they turn on groupings which involve cooperation among adult males. And it seems to me that this larger-than-family grouping—tribe, clan, pack, whatever you want to call it—must go far back in the history of the evolution of human behavior. Otherwise I don't see how man's unique characteristic—the development and accumulation of cultural traits—can be explained. The whole overlay of culture depends on an elaborate communication system, language, and I can most easily understand the development of this elaborate system as a response to a need for quick communication among many individuals involved in joint enterprises.

This gets us into a very old discussion as to whether the family is a primary, anciently basic human institution, or whether it is a secondary development arising gradually within hordes that were, to start with, sexually promiscuous. Today, I think, most psychologists and anthropologists regard the family as a primary, basic institution, carrying over directly from the family organization found so widely among the mammals. They depend on this, among other things, to explain the universality of some sort of incest taboo in all known human cultures.

I must confess that I have no alternative theory as to how incest taboos may have originated. But I belong to the minority group in thinking that the nuclear family, in man, is most probably a secondary development. This may make it more difficult to explain incest, but it seems to me to make it easier to explain a large number of other human characteristics. This is the sort of thing that can never be settled definitely, because we will never have eyewitnesses to the behavior of Peking men or Java men or the South African ape-men. We have to deal with inference, deduction, circumstantial evidence. Which makes the game of trying to explain the human species even more interesting.

14. The Human Species

> The more I see of uncivilized people, the better I think of human nature and the essential differences between civilized and savage men seem to disappear.
>
> —ALFRED RUSSEL WALLACE, in *My Life*

Linnaeus, who invented the system of cataloging nature that we still use (giving every animal and plant a name according to its genus and species) started this book with man—the genus *Homo* and the single species *sapiens,* though with a number of varieties recognizable by skin color and temperament. *Homo sapiens* we still are in the catalog of nature, a single species represented by nearly three billion specimens crawling everywhere over the continents and islands, floating everywhere over the surface of the seas. We are about four times as numerous now as in 1758, when Linnaeus published the definitive edition of his catalog. And lately we have been multiplying at a staggering rate, with a net gain of something like twenty-five million more human beings every year—a new population equivalent in size to the city of Detroit every month. Presently, if this continues, there will be no room for anything else on land—though there still may be space in the seas. This human population growth has been aptly compared with the wild growth of cancer cells, and perhaps carries within it, like a cancer, the doom of the biosphere—and the doom of man himself. How did this animal get into this peculiar position?

If we examine a pickled specimen of *Homo sapiens,* it doesn't look so very peculiar. It is curiously hairless, the brain is relatively large, the teeth small and unspecialized, and there are various changes in the skeleton related to the animal's habit of walking upright on its hind legs. It is clearly a primate, fitting in neatly with the monkeys and apes—really very similar to the apes, but perhaps differing enough to be put in a separate family, the Hominidae.

The primate order includes a rather diverse collection of mammals. Except for man, they are almost exclusively tropical, though one monkey occurs in Algeria and Morocco (with a colony on the Rock of Gibraltar), another ranges quite far north in Japan, and baboons range south of the tropic in Africa. With the exception of men and baboons, they are forest animals. None of the primates, except man, is really abundant or conspicuous, and most of them are shy and rarely seen. Macaques, of course, are common enough in some parts of India, baboons succeed in annoying South African farmers, and the tropical American howler monkeys in their remote forests make the biggest noise in the animal kingdom, but that is about the catalog of conspicuous achievement by nonhuman primates.

To go on with the classification business, the four ape types —gorilla, orang-utan, chimpanzee and gibbon—are put together in a family, the Pongidae. All of the endlessly varied African and Asiatic monkeys and baboons are put together in another family, the Cercopithecidae. These Old World monkeys differ in many basic items of anatomy from their New World counterparts and it is clear that the two have had independent evolutionary histories for a very long time. Any monkey that goes in for bright splashes of color on a naked face or buttocks is from the Old World—an emphasis on the buttocks is a sort of general characteristic of Old World monkeys. The commonest monkey in laboratories and zoos is one of these, the macaque or rhesus of India. When a psychologist talks about the behavior and intelligence of the monkey he means the Indian macaque, since generally this is the only monkey he knows.

The New World monkeys—classified as the family Cebidae —are an extraordinarily diversified group, both in looks and temperament, and they deserve a great deal more study than they have received. The trouble is that many of these monkeys don't do very well in cages; and since psychologists, as a rule, don't do very well in tropical forests, the psychologists and the monkeys rarely come in contact. When they do encounter each other, as in the case of C. R. Carpenter and the howler monkeys of Panama, the result is likely to be a very significant contribution to our knowledge of animal behavior.

We kept seven different kinds of these monkeys in the yellow fever laboratory at Villavicencio, and they were so fascinating that I had to keep reminding myself that I was studying yellow fever, not monkeys. The most striking thing was the distinctness, both in temperament and behavior, of each of these kinds of monkeys. If we ever find a neat, comparative way to measure "intelligence" (whatever that is) among animals, I'll bet that the Cebus, the traditional monkey of the organ grinders, will come out on top, with the chimpanzee as its only likely competitor.

The late Earnest Hooton of Harvard aptly called the Cebus "the monkey mechanic" because it can solve tool-using problems better than any ape—things like using a wire hook to get a short T-stick to knock down a long T-stick which could be used for knocking down a banana. The Cebus can even "make" tools in the sense of tearing off pieces of newspaper and rolling them to rake in some desired object.

I don't recommend Cebus as pets—they are too bright and too little interested in human values. They don't appreciate the human idea of discipline. They are given to violent dislikes—and they are apt to take an aversion to experimental psychologists. Heinrich Klüver, the gentlest of psychologists and the most successful with American monkeys, was never able to establish any rapport with his brightest Cebus. They seem also to get a great deal of pleasure out of annoying humans, which is understandable enough. I used to have a Cebus named Roberta, who was very good at this. She learned to snatch the glasses off the nose of a visitor and then run up into a tree, brandishing them. She never broke the glasses, but she did enjoy the reactions of the visitors. This same Cebus really hated the Indian macaques and when she had a chance would spend hours throwing stones at their cages, which made a continuing repair problem. She got into the animal house kitchen one night, and I have never seen a bigger mess than the one she left. Eggs formed the basic ingredient, splashed against the walls, the ceiling, everywhere; cabbages had been carefully dissected into small bits; everything had been opened and spilled. It must have been a glorious spree. But you can see the problems of having a Cebus about the house.

Roberta and I achieved a certain precarious understanding —which my wife said was based on the fact that we were both unpredictable—but I never had a feeling of shared affection with Roberta or with any other Cebus. For affection, I recommend the monkey acrobat, the spider monkey. Spider monkeys seem to love the whole world; when this is not reciprocated by man or animal, they simply bound off to some safe perch and scold with gentle disapproval. I suppose in their own forests, with their lithe bodies and beautiful muscular coordination, they are safe from everything except man with gun or bow. They can afford the rare forest luxury of gentleness and tolerance. But again they are not adapted to houses. I don't think I have ever seen a spider monkey intentionally break anything; but the world, to their mind, is one vast gymnasium. Lamps, curtains, light wires, are all interesting only for their acrobatic possibilities, which is hard both on the lighting system and on the monkeys. We twice tried giving spider monkeys their freedom around the laboratory, and each time they ended by electrocuting themselves.

But I started out to put man in perspective with the rest of the primates, not to tell stories about South American monkeys. The point is that the Cebidae, cut off from the rest of the world for dozens of millions of years in the South American forests, show a variety of possible trends in primate evolution—in anatomy, in behavior, and in temperament—and the implications of this have, I think, been very inadequately explored. There is the mechanical Cebus; the acrobatic spider monkey; the inquisitive squirrel monkey; the morose Callicebus, well-called the widow monkey; the clannishly communal howler monkey which no one so far has succeeded in bringing into a laboratory; the woolly monkey, an animated and independent-minded teddybear that still likes to be cuddled; the owl-faced night monkey, which makes the best pet of all, especially for people with insomnia. And then there is a whole group of marmosets—tiny, fierce little things which are generally classified in a separate family from the more conspicuous Cebidae.

To complete the survey of the primates, we should say something about a curious, rare little creature called the tarsier which lives in the East Indies. It is of great interest to zoologists because it is the only survivor of a kind of primate common in the Eocene, sixty million years ago, that may have been ancestral to all the others. And then there are all sorts of lemurs, mostly living on Madagascar where, protected from the evolutionary competition of the continents, they have developed a whole array of special forms.

What does this tell us about man? Not much, directly. The more we know about the anatomy and behavior of these various primates, the richer background we have for looking at the human experience. But clearly the living primates do not represent a surviving record of a sequence from tarsier to monkey to ape to man. In anatomy, the apes are most like men. In tool-using and in group organization, it seems to me that the distantly related New World monkeys are more human-like. But men, apes, and monkeys have all been evolving at the same time, changing and modifying the characteristics of their remote common ancestors. Evolution in man may have gone on at a faster rate than in apes, characteristics may have persisted in one kind of animal and not in another, but this is not easy to judge. The method of comparing different kinds of living organisms is not, in itself, an adequate basis for reconstructing a plausible evolutionary story, however rich it may be in suggestions of plots for that story.

Our surest way of reconstructing the past is through looking at the historical record—through looking at the documents, at the fossils, that have survived. Unfortunately, the record is extremely incomplete, composed of a small and haphazard collection of fragments, sometimes difficult to date and always difficult to interpret. We have teeth, jaw fragments, skulls, sometimes fairly complete skeletons, from various kinds of manlike animals that have lived at different places and at different times during the last million years. And we have a wide variety of tools, mostly of stone, chipped or polished to shape them for some particular purpose, from many parts of the world. But how do we put flesh on these bones? How, from a chipped flint, do we deduce a way of life?

The fossils, whether bones from the body or tools from the culture, are clues which must be interpreted. Someone has suggested that archaeologists study police manuals because the problem of the detective and the problem of the archaeologist is the same. Nature has no more intention of leaving a record than the criminal does and the surviving evidence is purely the consequence of chance. The evidence is necessarily circumstantial, since we have no eyewitness and no possibility of finding one. We have only the faint and inadequate clues. But these are facts which must be accounted for, must be explained. The clues serve as take-off points for our reconstructions, and this is where we have to fall back on our knowledge of various kinds of living things, on our knowledge of the behavior of living animals and of the characteristics of surviving human cultures. In themselves, these things do not form a historical sequence, but they may give us, a bit here and a bit there, the evidence needed for building a possible historical sequence.

The problem of explaining the development of man is the problem of explaining the development of the human brain and all the things related to this brain—learning, thought, language, social behavior, culture. As I remarked before, if you look at human anatomy or physiology there doesn't seem to be anything very peculiar about it; but if you look at human behavior, at what men do, it seems to be very peculiar indeed. Man is quite different from anything else in nature, and we have to face the question of how this difference can be explained by natural processes.

Many people—many scientists—have felt that man cannot be explained by natural processes and have fallen back on the supernatural. This seems to me no help at all, since the operation of special supernatural agencies to explain man creates more problems that it solves. If nature is orderly, it ought to be orderly all the way through and I can't see what is gained by supposing some special, miraculous upset in the order back about the middle of Pleistocene times.

It is interesting, in this connection, to look at the difference in opinion between Charles Darwin and Alfred Russel Wallace. Wallace hit upon the idea of evolution through natural selection quite independently of Darwin and the idea was first presented to the world in papers written jointly and read before the Linnaean Society on June 30, 1858—a momentous date in the history of ideas.

Darwin, for the rest of his life, felt that the theory served to explain the whole living world, including man and his institutions and ideas. Wallace, perhaps an equally great naturalist, but with a much deeper knowledge of primitive peoples, came to feel that the human mind could not be explained by natural selection or by any other evolutionary process. Wallace thought that the gap between the animal brain and the human brain was so wide that it could not be bridged by imaginable transitions. I don't know of any biologist or anthropologist today who would agree with Wallace; yet, as Loren

Eiseley has pointed out in *Darwin's Century,* Wallace in many ways had a more modern understanding of the nature of man than Darwin did.

At that time, the only human fossils that had been discovered were some Neanderthal remains, and these were not recognized as fossil remains of an extinct human type. The evidence for human evolution, then, was entirely comparative, based on the similarities between apes and men. For Darwin, the various savage tribes represented an evolutionary sequence of sorts between the ape condition and the condition of civilized man. Wallace, with his years of travel and living with primitive tribes in South America and Malaya, felt that these people were no different from him in mental ability, moral sense, language development, physical capability, or anything else except material goods—except what we today would call culture. For Wallace, the mind of man was essentially the same everywhere, among all peoples. The Dyak, had he been born in England, might have made a brilliant record at Oxford; while the Englishman, had he been born in the hills of Borneo, would inevitably have followed the Dyak way of life. How could one explain this human mind, with all of its wonderful potentialities, in terms of slow development through selective forces operating in a tropical forest? What, in this environment, would lead to the development of an animal capable of making and running a steam engine or of composing a symphony? It is understandable that Wallace retreated into mysticism.

The difficulty, I think, lies in looking at tools, at culture, as the product of the human brain. For me, the difficulty disappears if we turn the proposition around and look at the human brain as a product, a consequence, of the use of tools, the development of culture. And this sequence is increasingly supported by the accumulating bits of evidence from the fossil record. Some animal has been using tools, shaping rocks, for a very long time, because such stones turn up commonly in deposits that can be dated back to the beginning of the Pleistocene, about a million years ago. And, looking back over the record of this million years, one can see a gradual increase in the skill with which the stones were shaped, and a gradual increase in the diversity of kinds of tools that were made. Improvement and diversity come with increasing speed in the last fifty thousand years, merging into the dizzy rate of change observable in the five thousand years or so of conventional human history.

We have many more fossil tools than we have fossil bones of men or manlike animals, and we only rarely find tools and bones in close association, so that it is difficult to be sure what kind of animal made what kind of tool. Increasingly, however, anthropologists are coming to believe that man in the strict sense, *Homo sapiens,* may be a quite recent evolutionary product, with a history extending back no more than, say, fifty or a hundred thousand years. Most of the Pleistocene tools, then, were made by pre-*sapiens* animals. In the case of Pekin man, *Sinanthropus*—a pretty low-browed fellow—crude stone tools have been found in association with the skeletons.

The greatest interest attaches to the Australopithecines, the South African ape-men, because they represent a very primitive manlike form living about the beginning of the Pleistocene. Did they have fire and did they make tools? The evidence is indirect. These animals did have greatly reduced canine teeth, as do all of the human line, in contrast with the great apes. It can be plausibly argued that the canine teeth would not become reduced in size except in a tool-using animal, which would no longer have need for the big canines.

If we imagine an animal without much more intelligence than, say, a chimpanzee, coming to depend on tools—on sticks and clubs and rocks—we can see that the whole action of natural selection would change. The individuals with the greatest ability to make, and use, tools would be favored. To understand man at all we have to presume, at this beginning of humanness, a social animal with some taste for meat, built in such a way that the hands could be used for handling tools. A cooperating social group with tools could afford to develop

—or retain—man's puny physique and generally unspecialized body characteristics. Evolution would turn, not on brawn, but on brains.

It seems that for a long time the chief enemies of men have been men: the force of natural selection has depended on competition within the human or pre-human groups. These socialized, tool-using animals must fairly soon have become relatively safe from the attacks of lions, crocodiles and similar predators. But they also fairly soon took to killing each other on a scale that has no parallel elsewhere in the animal kingdom. If you examine the known human or pre-human fossils with an eye to determining the cause of death, it turns out in a surprisingly large number of cases that the individual, quite clearly, has been murdered. Surviving skulls show that they were pierced by spears or bashed by clubs. Long bones often are split open. No animal except man would be able to split a bone, and the only conceivable purpose in splitting a bone would be to get at the tasty marrow. This, then, shows that both murder and cannibalism were ancient human practices.

I can understand this only in terms of territorial behavior. Fighting over territory is common among many kinds of vertebrates. The loser is rarely, if ever killed; he is simply driven off. Konrad Lorenz explains this in terms of biological evolution: that as weapons like teeth and claws developed that would enable an individual to kill another member of his own species, inhibiting behavior developed that prevented fighting from being carried to the killing point. With man, however, weapons are the product of cultural, rather than biological, evolution, and inhibiting behavior has simply not kept pace with weapon development. But whatever the evolutionary explanation, the result is clear; men have been killing each other for a long time.

If we look at Carpenter's howler monkeys, with their strong clan and territory organization, we can see how this might have developed. The monkey clans, when they meet on their territorial borders, simply howl at each other until one group or the other retreats. Monkeys not belonging to the clan are clearly strangers, outsiders, viewed with great suspicion. The howling is harmless enough; but if these monkeys had spears and clubs, the result of the border squabbles might be quite different. With primitive, food-gathering people today, we can see a clan and territory organization quite similar to that of Carpenter's monkeys. And inter-tribal squabbling quite often has deadly consequences. In cannibalism, the stranger from some foreign tribe is clearly not regarded as human, as a member of the group; and if you kill him, you might as well eat him. There is no point in letting good meat go to waste, especially when it is so hard to come by. Which makes humans pretty "inhuman" and "unnatural."

Given this sort of situation, survival and successful reproduction would come to depend on improved ability to make and use tools, and on improved cohesiveness and communication within the effective group—tribe, clan, or whatever you want to call it. We would thus have continuing selective pressure for what we look at as the distinctively human characteristics. The individuals and tribes making the most effective weapons and using them most cleverly would win out. This, at first, would involve biological traits: brain, muscular coordination, speech development and behavioral patterns leading to group solidarity. It would also put a premium on learning, on ability to modify behavior according to the circumstances.

With the development of speech and the possibility of accumulating information (and misinformation), this same intra-human, inter-tribal, competition would guide the development of cultural evolution. From the record of fossil bones and surviving tools, it looks as though once man the animal reached the point of effectively developing cultural traditions, cultural evolution, with its more rapid pace, took over, until now it is the major force governing human characteristics and differences.

But it was a long time before biological evolution was swamped by cultural evolution. The low-browed pre-humans that started using tools were under constant selective pressure to develop the biological equipment—brains, easy upright posture giving freedom to the hands, instinctive and anatomical traits allowing speech development—necessary for the mastery of the tools. Which is why it seems to me that the human brain is more easily understood as a consequence of culture, of tool-using, than as something that had to be developed before culture could be acquired.

But this biological equipment, way back in the Stone Age, reached the point that allowed all of man's subsequent cultural developments to take place. The skills that were required for survival under Stone Age conditions were, it turned out, the skills behind all of man's subsequent bizarre achievement. This is what Darwin failed to understand because he never clearly saw the difference between man as a biological animal and man as a consequence of cultural history. Wallace saw that modern man, biologically, was the same everywhere, and that his differences in accomplishment were the products of his differing cultures, but he failed to see how this potentiality for cultural development could have arisen as a consequence of organic evolution—hence his retreat into mysticism.

It takes a great deal of skill to do a good job of chipping a flint—some of our brightest people have tried to master the art with only indifferent success. Once you have acquired the brains and coordination of eye and hand necessary for shaping a flint, or for carrying out some similar Stone Age activity, you have all the biological equipment necessary for making and flying an airplane or for carrying out any of the other activities of modern civilization. The Dyak of Borneo, as Wallace saw, has the same biological equipment as the Oxford don—though the two use this equipment in different ways, for different ends, in accordance with different cultural traditions.

The difference between man and other animals, when we look about us, seems tremendous. Man's mind seems to differ fundamentally from anything else we can see in nature. But this, as the psychologist Harry Harlow has pointed out, is because we fail to distinguish between capability and achievement. Man's achievement, certainly, is extraordinary when compared with that of any other animal, but this does not necessarily depend on a correspondingly great shift in biological make-up, in capability. As Harlow notes, "The fledgling swallow a few days before it can fly differs little in anatomical and physiological capacity from the swallow capable of sustained flight, but from the point of view of achievement the two are separated by what appears to be an abysmal gulf."

Students of comparative psychology, like Harlow, find a gradual increase in learning ability among the mammals, from rats, to dogs, to monkeys and apes and man. There is no abrupt shift in ability to solve the kind of learning and associative problems that psychologists love to figure out. Adult monkeys and apes, in fact, may be better at solving some kinds of complex problems than human children—and human adults don't necessarily turn out to be very clever when in the hands of experimental psychologists.

Then we have the several cases in which baby chimpanzees have been raised in human families. The chimps always for a while outstrip human babies in accomplishment, but presently fall back when language communication takes ascendancy. Still, the ability of these chimps to cope with the human situation, with human gadgets like tableware and toilets, and with human customs, remains extraordinary. Chimps obviously have all sorts of potentialities never realized in their native forests, potentialities that find expression only when the animal is subjected to the strange environment developed by human culture. This, of course, is true to some degree of many animals when kept by men as pets. The behavior of many animals besides men may be modified by the influence of culture, but only man has developed culture. Why?

I have argued that the biological change that would get culture started may have been relatively small. But what sort of a change would it be? We get one clue from the chimps raised in human families: they drop behind at the point where language becomes dominant in human development. They "understand" and respond to a considerable variety of word sounds, but they make no effort to imitate. There doesn't seem to be any anatomical reason why they couldn't at least produce some speech sounds. The whole impulse to babble and imitate, so prominent in the human baby, is missing.

Sound production is very basic to human nature, to human sociability. There are occasional strong and silent individuals, but most of us live in a constant welter of sound exchange, not necessarily involving communication; we are uncomfortable in groups if we are silent, so we constantly chat, constantly engage in "phatic communion." This desire to make noise, this pleasure in the interchange of sounds, this tendency to mimic and repeat and give meaning to vocalizations, may have started from some rather small shift in the genetic make-up of the pre-human stock—a slight shift with enormous consequences.

We know remarkably little about the genetics of human evolution, about the kinds of mutations that might have led to the kind of physical animal we recognize as contemporary man. We know remarkably little, for that matter, about our present genetics—about the hereditary controls that shape normal human beings. We can't experiment with laboratory colonies of people, which makes it difficult to get such information. We have to fall back on abnormalities which can be easily traced in genealogies, or on traits like blood groups that have relatively simple genetic controls. Again the fossil record isn't of much help because it is so very incomplete and because it consists entirely of bones, which tell us nothing about things like skin or hair; or of tools, which tell us about behavior only in the most indirect way.

If we look at contemporary men and contemporary apes, however, it is interesting to note that men, in many ways, are much more like baby apes than like adult apes. This is true of the proportion of the brain to the rest of the body; of the relatively hairless condition of the skin; of the failure of the human skull to develop heavy brow ridges. The whole human head and face, in fact, is definitely infantile when compared with the heads of other primates. Man also has a very long gestation period; and after birth is extraordinarily slow about growing up, about achieving independence as an individual, and sexual maturity.

A Dutchman, L. Bolk, noted all of these things in 1926, and suggested that man might be a sort of fetalized ape, a kind of ape that has lost the capacity to grow up. Various cases are known in insects, salamanders and other animal groups in which larval forms become reproductively functional, and the normal, adult state is never reached. This is called *neoteny* and it may have been an important factor at various crucial points in the history of animal evolution.

I must say I like the idea of man being an ape that has lost the ability to grow up. The prolonging of the period of infantile and juvenile growth and the shift in the onset of sexual functioning in relation to other physical developments could be easily enough explained by minute changes in the functioning of various endocrine glands, which in turn are directly under genetic control. Again, a slight change in biological make-up could have disproportionate consequences.

Many of the more reassuring aspects of human nature can be explained in terms of this persisting childishness. The young of many animals show a great development of play activities—which seem to me not to have received enough attention from the students of animal behavior. Play—however defined or explained—is clearly important not only in childhood but all through human life. I wonder whether many of the most characteristic of human activities may not have come out of play behavior—magic, ritual, art, and even science.

This, of course, leaves many aspects of man unexplained—among other things, his continuous sexuality (though from watching puppies, I sometimes think maybe this, too, is juvenile). This continuous sexuality—whether in the indulgence or the frustration—is, as the psychologists all tell us, fundamental to the understanding of human social behavior. Sexual regularities often break down in the animals that man has domesticated, and many authors have regarded man's sexual behavior as a consequence of a similar process—a process of "self-domestication." If this were true, human sexuality could be regarded as a consequence of cultural development rather than a cause—a consequence of man's relative security from non-human enemies, of his relatively safe and continuously available food supply; of, in short, his escape from the rigors that control the life cycles of other animal populations. But this whole business of domestication is such an interesting thing, biologically, that I think it warrants examination in a separate chapter.

15. On Being Domesticated

Our civilization still rests, and will continue to rest, on the discoveries made by peoples for the most part unknown to history. Historic man has added no plant or animal of major importance to the domesticated forms on which he depends. He has learned lately to explain a good part of the mechanisms of selection, but the arts thereof are immemorial and represent an achievement that merits our respect and attention.

—CARL SAUER,
in *Agricultural Origins and Dispersals*

Man—in his present numbers at least—has come to depend on his domesticated animals and cultivated plants for food, and he could not survive without them. And the cultivated or domesticated animals and plants—we might lump them together under the elegant word *cultigens* —have come to depend on man for survival, in varying degree. Corn (maize) is a nice example of the extreme case. How could corn, as we know it, possibly survive without man to husk the ears, remove the kernels, and plant them? How, on the other hand, could the Indians of the Mexican highlands survive without their corn, without their tortillas? This, in the jargon of ecology, is a case of mutualistic symbiosis, though it is not often called that.

We can find parallels among other sorts of organisms. The closest are the fungus-cultivating ants and termites. There is a group of ant species in the American tropics, extending into the southwestern United States, that have come to depend for food on fungi that they cultivate in their nests. The foraging worker ants cut out roughly circular bits of leaves from various favored species of shrubs and trees; these are carried back to the nest to be deposited in special chambers to form a substrate on which spores of a particular kind of fungus, a mushroom, are planted. A caste of very small-sized workers is specialized to care for these fungus gardens, destroying alien "weeds." The fungus forms the sole food of the ant colony. Winged female ants, leaving the nests to mate and found new colonies, carry a packet of spores of the fungus in a cheek pouch so that a new garden can be started in the new colony. The particular species of mushrooms cultivated by these ants have never been found growing anywhere except in the ant nests, so that they presumably have lost the ability to survive except when cultivated by the ants. The ants, in their turn, have become completely dependent on the fungus.

The ants apparently instinctively know which kinds of leaves are best suited for growing their fungi—at least they are highly selective and rapidly denude favored trees. Since they often attack orchard trees or trees used for shading coffee, the ants are a major pest from man's point of view. I remember a nice demonstration that I once saw in Honduras of the adaptability of the instincts of these ants. A young wife, newly arrived from England, decided to bake some bread as part of her effort to make things homey in this strange place. She put her pans of dough out on the back porch to rise. When she went back a few hours later, she found a busy column of leaf-cutting ants carrying off the last of her dough! The ants never could have encountered any material of this sort before, but they must have been aware, through some instinct, that it was a fine substance for growing their fungus. The English wife, who didn't like tropical insects anyway, failed to appreciate the interest of this situation.

Ants hardly differing from kinds living today have been found in Baltic amber, fossilized fifty or sixty million years ago. There is every reason to believe that the complicated habits of ants, including behavior like fungus cultivation, have developed through these unimaginable stretches of time. They thus can hardly be compared with human activities, which have developed with extreme speed. Man's cultivation of plants and domestication of animals has a known history of something over five thousand years, and the fumbling, guessed-at beginnings can hardly go back more than five thousand years more. The difference, of course, is that the human behavior is based on learning, culturally transmitted, while the ant behavior is based on instinct, biologically transmitted.

We can assume that the human biological constitution has not changed during the ten thousand years or so that these cultivating habits have developed. But the biological constitution of the cultigens has changed—and this, to me, is an extraordinary thing. We know of nothing anywhere else in nature comparable with the speed of biological change in most of the plants and animals presently cultivated by man. We can surely learn a great deal from this. We have learned much—for one thing, Darwin and Wallace got the idea of natural selection from developing an analogy with man's artificial selection. But somehow I still feel we don't quite appreciate what a peculiar phenomenon we have before us in man as an agent of biological change. I suspect we still have many lessons to be learned from this about the operation of evolution.

Before going into the special case of domestication, let's try to reconstruct something of the history of man's changing relations with other members of the biological community. The primates ancestral to man were hunters. This is clear from the earliest tools, from animal remains, from any attempt to explain the habits and disposition of modern man—civilized or not. Meat-eating is an unusual habit among the primates. The monkeys and apes generally are either vegetarian, insectivorous, or omnivorous in the sense of eating anything they can get hold of, whether fruits or worms. Man, too, has probably always been relatively omnivorous, scrounging all sorts of tidbits from his environment. But to have this as the total food supply is quite different from having it as a supplement to the rich supplies of protein available if one can catch and kill the herbivores, the teeming grazing animals.

This taste for meat, this shift to predation, however it got started in the human line, had many implications. Primate social aggregations, whether of howler monkeys or chimpanzees, are primarily defensive: there is a great advantage for individual survival in belonging to a group, as long as the group doesn't become so large that finding adequate food supplies becomes a problem. But the advantages multiply with the shift from defense to aggression, especially for an animal without great physical prowess and without special biological equipment in teeth and claws and muscle. It is only by "ganging up" that animals like the pre-human primates could take up the predatory way of life. And the tendency to gang up, to develop social cohesiveness including skill in communication, would be strengthened by the continuing development of predatory habits. It seems quite likely that this hunting-pack stage has left a deep impress on our psychology.

The relations of this predatory, pre-human, primate with the biological community would be similar to those of wolves or coyotes. The pre-humans would undoubtedly sometimes be killed and eaten by other predators, like the big cats; but the pack organization would be a considerable protection, making this unimportant except for old or weak individuals or straying young. The pre-humans, like other animals, would have a collection of parasites. Various internal parasites like some of the intestinal worms and the protozoan causing malaria probably evolved right along with man. There would also be external parasites, like a special kind of body louse; and, along with other mammals, these pre-humans would serve as food for many kinds of ticks, mites, mosquitoes and the like.

But the pre-humans would be more important as predators than as prey. As they gained in social hunting skill and in skill in making weapons and traps, their predatory role in the community would become more and more important. It looks as though man presently became a good enough hunter to at least be a contributing agent in the extinction of some kinds of prey, like the mammoths and ground sloths. And, as I have remarked, man early developed the curious custom of killing other men, which must have had important evolutionary consequences.

When humans or pre-humans started building shelters for long term occupation, many other kinds of animals found these shelters convenient for hiding or foraging, which resulted in a new set of relations with the biological community. Bird nests, prairie dog burrows, all such animal homes attract a variety of uninvited guests. The nests of social insects like ants and termites have become the habitat of a whole special fauna of beetles, flies and the like and the members of this fauna are greatly modified both in habit and appearance for life in this environment. Some of these "guests" cause direct damage to their hosts as parasites or predators; others simply take advantage of the special shelter or food situation without causing appreciable harm. Biologists call these associates of ants and termites *inquilines* and this seems to me an apt term for the uninvited companions that man started accumulating as soon as he began making shelters.

I don't know that anyone has made a special study of the human inquilines, though I think such a study would be very interesting. Any shelter in the tropical forest or savanna, where man presumably started making shelters, soon accumulates its own fauna: cockroaches, scorpions, lizards, ants; a variety of animals sometimes of species not otherwise common. I would call the bedbug an inquiline, one so adapted to life in human shelters that it must have been associated with man for a long time; the closest relatives of bedbugs inhabit places like bird nests. The house mouse and certain species of rats also probably discovered the advantages of associating themselves with man a long time ago.

These inquilines in many cases could well be called domestic, but they could hardly be called domesticated. Man's first domesticated animal, as far as we know, was the dog. We may have a transition here between the inquiline and the domesticate. The man-dog relation goes back for a very long time and one can plausibly argue about whether man adopted the dog or the dog adopted man. The association might have started by dogs, jackal-like, lingering around human kills for the offal, developing gradually to the point of sharing in the actual hunt. The dog's better sense of smell and sense of hearing supplement nicely man's better vision and greater ingenuity. The only trouble with this theory of shared advantages in hunting is that it looks as though hunting dogs were a rather late and special development.

A great deal has been written about the origin of the dog, but there is still no clear agreement among the experts as to whether the domestic dog is descended from one of the wild canines still existing, whether it represents a mixture of several wild species, or whether it originated from a species now extinct in the wild state. It seems most likely to me that the ancestral dog represented some wild species now extinct—perhaps a species resembling the Australian dingo—though it is clear enough that our modern dogs include also strains resulting from crosses with various species of wolves and wild dogs in different parts of the world.

It is curious how little we know for sure about the origins of any of our important domesticated animals and plants, despite the vast amount of work that has gone into the study of the question. When these animals and plants emerge on the historic scene depicted on monuments, preserved in tombs, or described in writings, they are already highly modified, difficult to identify with any particular wild species. The domestications took place while man was still living in close association with the biological community. We can document the exchanges within historic times—the introduction of citrus fruits into Europe from the East or of potatoes and maize from the West, the transportation of the whole catalog of Old World crops to the New World and vice versa. But as we try to track down the beginnings of any particular cultivation, we find ourselves lost in the circumstances of some preliterate cultural area.

After the dog, even the sequence becomes uncertain. The

old idea that pastoral cultures, depending on the herding of grazing animals, developed from hunting cultures and then passed on into the agricultural stage with planted crops and settled villages, has long been abandoned. The food-gathering, hunting, pastoral, fishing and crop-raising cultures that we find in different parts of the world today can no more be arranged in an evolutionary sequence than can the different living monkeys, apes and men. The surviving cultures, like the surviving animals, are the products of their particular differing evolutionary histories. They can only give us clues as to what might have happened, which we can check against the fragmentary historical record, and against our own analysis of the logic of events.

Carl Sauer has made an interesting analysis of the conditions under which agriculture may have started. He thinks, first of all, that cultivation likely began among people who already had adequate food resources: "People living in the shadow of famine do not have the means or time to undertake the slow and leisurely experimental steps out of which a better and different food supply is to develop in a somewhat distant future." And, he adds, "the saying that necessity is the mother of invention largely is not true."

Sauer also considers it likely that agriculture got started among sedentary people, rather than among roving hunters. Pre-agricultural people with a settled way of life and with adequate resources were, most likely, fishing folk living in a mild climate, probably in forested country. Grassland is very difficult to cultivate with primitive implements—forest is easier to deal with because the trees can be cut or girdled with stone axes, and burned. The most primitive agriculture still is tropical forest agriculture.

Planting may well have started with root crops. Food-gatherers have been digging roots for a long time, and the idea of putting back some of the root or stem to make a new plant seems easier to come by than the idea of harvesting and planting seeds. Root crops are particularly characteristic of the tropics—yams, taro, cassava and the like. Southeast Asia,

Sauer thinks, is the most likely place in which to look for the origins of agriculture in the Old World: a place where there is a wide variety of animals and plants to be experimented with. Southeast Asia, further, seems to be the original home of a surprising number of domesticated plants and animals. Unfortunately, we know little about the archaeology of the region, since Old World archaeologists tend to live in Europe, which was on the fringe of things during most of the course of human development.

Primitive man—Paleolithic and early Neolithic man—lived close to nature, was a part of nature, in ways that are hard for us to imagine. He had to know his environment intimately, had to make correct observations and deductions, in order to survive. And he had an immensity of time during which he could accumulate knowledge which would become traditional with his tribe and culture. When we look at all of the foods, medicines, drugs, arrow poisons, fish poisons, fibers—sometimes depending on complicated methods of preparation—the accomplishment seems incredible. But this early man, at least in some kinds of cultures, lived in daily association with the plants and had a great many thousands of years in which to discover their properties.

The intimate relations possible between man and flora were impressed on me during the summer I spent on the Micronesian atoll of Ifaluk. The flora on this oceanic island was, to be sure, very limited. We found only 119 species of ferns and seed plants, and I calculated that about a third of these were purposeful introductions, another third accidental introductions, and the remainder truly indigenous local species. The people had distinctive names for all but six of these species—three of the unnamed were grasses that it would take a botanist to distinguish, and the other three apparently were recent accidental introductions of weeds.

There were only 14 of the 119 species for which our informants said there was no use. It is perhaps significant that the most frequent "use" was in the preparation of medicines of one sort or another—51 species were involved in medicine, as compared with 39 species used in garlands for personal ornament, 29 species used in construction of various kinds, and 26 species considered to have food value (the same plant, of course, frequently had multiple uses).

The people of Ifaluk are skilled agriculturists, living in a very special sort of environment, so that they can hardly be compared with a Stone Age people on the verge of developing cultivation. The Micronesians and Polynesians are also unusual in the importance they give to flowers as personal ornaments —though man has been given to ornamenting himself for magical or esthetic reasons for quite a while. Some students of the subject think that plants used for face or body paint may have been the first cultigens. Ornament is certainly significant, then, and so is medicine, in the development of cultivation.

Interestingly enough, there is a growing conviction among many students that the beginnings of animal domestication may have turned on magic rather than on food. In many cultures today, poultry are kept for divination—and for cock-fighting —rather than for food; and pigs may originally have had greater ceremonial than nutritional importance. Remnants of pig cults were found in many parts of the Mediterranean as late as classical times. The opposite, the powerful taboo on pigs in some religions, is most easily understood as a rejection of the heathenish ways of despised cults—and to be despised, the cults have to exist.

The dog, pig, fowl, duck and goose are all household animals, and all may have been first domesticated in southeast Asia in association with people depending on root crops or other kinds of vegetative propagation in their agriculture. The herd animals, on the other hand, first appeared in southwest Asia in association with seed agriculture, with grains. The herd animals are such obviously useful additions to man's resources that many attempts have been made to explain their origin in rational, practical terms. It is true that cattle do not seem very practical in a culture where prestige turns on ownership, so that the "sacred cows" to an outsider look like cultural handicaps. And anyone reading history gets the idea that the horse wasn't of much use until the Middle Ages except in war, where there is always the suspicion that the war stories were written by frustrated cavalrymen or charioteers. But donkeys, milk cows, goats and wool-bearing sheep look practical enough. Yet the whole herding complex may be most easily understood in terms of irrational origins.

Why, with the host of species of antelopes, gazelles, and deer easily available and easily tamed, often kept as pets, did man concentrate on domesticating the particular animals that he did? As Sauer remarks, "One might say that animals were chosen for domestication that were not easy to take, which were not common, and which were difficult to make gentle—the wild mountain goats and sheep that avoid the vicinity of man, the formidable wild cattle and buffalo." All of these, in the beginning, make more sense in ceremonial than in practical terms. All of the herd animals were, at one time, milked—a curious practice that is to this day rejected with horror by many cultures. The German geographer, Eduard Hahn, many years ago suggested that this might have started in connection with the widespread cult of the mother-goddess in the dim beginnings of the Near Eastern and Mediterranean civilizations. And cattle, of course, drew the phallic plow that quickened the earth.

But this is wandering far from biology. Whatever the explanation, practical or ceremonial, man started to enter into a special relationship with a whole series of animals and plants perhaps ten thousand years ago in the Old World, somewhat more recently in the New. These animals and plants came to be greatly modified in the course of the passing centuries, came to form a special, man-centered biotic association. Man and these associates started the process of remaking the biosphere, a process that contains at an ever accelerating rate. This is of direct biological concern.

Man is cosmopolitan and a few of the members of the human association have accompanied man everywhere or almost everywhere—dogs, lice and fleas come to mind. Some domesticated animals have achieved very wide distribution—chickens, pigs, cattle, horses. In general climatic tolerances are wider for cultigens, animal or plant, than for corresponding wild species, in large part because the genetic variability is greater in cultigens. There are large differences, however, between the tropical and the temperate associations. In modern times geographical barriers to the spread of human associates have largely broken down because of the development of man's transportation systems, so that particular species—whether cultigens, inquilines, parasites or what—are liable to be found wherever conditions are appropriate, owing to either purposeful or accidental transport. But the distribution of many of the cultigens is also restricted by cultural factors. People tend to be conservative in their food habits and food crops in climatically similar parts of the world may be quite different for cultural reasons: basic crops which have become widely spread in modern times, like wheat, maize, rice and potatoes, are relatively few.

When one looks at the face of the land today, the changes brought about through human agency, direct or indirect, are very impressive. But if these changes are examined in greater detail, another impression develops: they are in large part temporary, reversible, and dependent on continuing human intervention. The whole biological complex centering on man shows a continuing dependence on human activity. The transience is particularly striking in the few places where civilized man, for some reason, has retreated: the rain forest, apparently undisturbed and unaltered, has taken over the great Mayan cities of Peten and the Khmer cities of Cambodia. The monuments of dead stone are there but the wounds made by men in the living communities have been healed. The crops, the weeds, the fields, the whole living entourage that depended on men has disappeared with hardly a trace, and we again have "primeval" forest.

If we examine the roadside plants on any continent, we find all sorts of things that have escaped from cultivation, or that have been accidentally introduced by human agency, growing along with local species. But when we move away from the roadside into undisturbed forest, the foreigners largely disappear. The effect of introduction is even more striking on islands. A student of the New Zealand flora found 603 species of plants that could be called naturalized, aliens that had successfully established themselves in this new environment. But when he started to examine these in greater detail, he found that most of them were growing only along roadsides, in abandoned fields, or in other situations drastically altered by man. Only 48 of the 603 species had been able to penetrate the indigenous biological communities where they were in direct competition with local species. Yet New Zealand, so long isolated from outside biological invasion, would appear particularly open to outside influence.

We read about wild dogs, hogs, horses, goats and the like, but when we look around the world, it is clear that domestic animals have rarely gone wild and become undomesticated—and then mostly on islands with few or no competing native mammals. Even man's attempts to move wild animals from one place to another have failed more often than not: rabbits in Australia and starlings in North America are exceptional successes which came only after repeated failures.

I would draw several conclusions from these observations. One is that the process of domestication is not readily reversible. Another is that established biological communities—the climax communities of the ecologists—are not readily penetrated by new species of animals and plants. A third is that there is something peculiar about the whole complex of organisms that thrive in association with human activity.

When man started cutting trees, making shelters and clearings, he started creating a rather unusual sort of situation in nature. Open soil, land where plants could grow but don't, is rare. The commonest forms of vacant land are the sand bars, mud banks and open spaces left by changing river systems. Forest fires set by lightning, landslides, and volcanic eruptions also suddenly create new situations. But for the most part environmental change in nature is slow, related to glacial retreat, lake sedimentation, or gradual shift in climate.

When man started creating open situations, a special group of plants moved in on him—the sort of thing we call weeds. Weeds seem tough and vigorous to us, when we try to get them out of our garden or lawn, but in the competition of the natural community they stand little chance. In the pre-Neolithic world they were rare, perhaps on the verge of extinction; man provided them with a saving habitat, and they have been depending on him ever since. Now many of our weeds and crop plants are closely related and things that are now crops were once weeds and vice versa. The extension of human clearing activities brought these open plants, from widely separated situations, into contact, and the previously separated species often hybridized. In such situations the ancestral species often disappeared in the swarms of hybrids, leaving us with puzzles which the geneticists are only slowly working out. Thus the weeds of our gardens have adopted man just as surely as the rats, mice and cockroaches of our houses. To some degree at least many of our cultigens may have domesticated themselves. Barley, we are pretty sure, started as a weed in wheat.

It is sometimes said that man himself is a "self-domesticated" animal. This easily starts arguments as to what is meant by domestication—arguments which are not very fruitful. But beneath the vocabulary problem there are certain realities that seem worth thinking about, certain similar tendencies among the diverse organisms that go to make up the human association. These similar tendencies, when we start to analyze them, are a consequence of the conditions of life under domestication—or, more broadly, of the conditions of life under the influence of human culture.

The human associates have, to a considerable degree, escaped from the usual complex system of checks and balances of the biological community. Natural selection is replaced by artificial selection. Concealing coloration, fleetness for escape, fierceness for defense, all lose their meaning. The pressures for conformity, for maintaining the species in the narrow niche in which it has successfully adapted, are either removed or utterly changed. Aberrant individuals that would be quickly eliminated in nature are tolerated and protected, or perhaps prized and pampered, because of some trait appealing to man.

The domesticates have, to a large degree, a protected and especially provided food supply. Adaptations to the changing seasons lose their importance. Dispersal mechanisms are not needed because transportation will be provided by man—chickens and silk moths equally lose the ability to fly. Individuals of the same kind are herded together so that elaborate systems for species recognition, for the timing and controlling of sexual behavior, are not needed. The conditions of life, in short, are greatly altered in many different ways.

One general result is a great increase in the genetic variability of a species under domestication. This is partly because mutations that would be eliminated in nature may survive; partly because the hybridization of different strains and even species is fostered by man's breaking down the usual barriers between populations. This increase of variation in the material available for selection—artificial or natural—is one of the reasons why evolutionary change proceeds at a faster rate within the human association than outside it.

If we look at the human species, we see many things that remind us of the characteristics of domesticated animals. The species shows a great deal of variation. Much of this is geographical, comparable with the geographical variation of many wild animals. But the extremes of human racial traits are greater than those of any wild animal I can think of. And in many parts of the world we can see evidence of the geographical variation being swamped by intermixture and hybridization —a situation like that found with many of the human associates. Some human populations look fairly uniform, but others show great individual variability in size, skin color, hair-form and physique. Blondness and blackness, which dominate certain human populations, are otherwise. with rare exceptions, found only in domesticated species. One wonders, comparing skin pigment in man and his domesticated animals, why spotted human varieties have not turned up.

The range in human size, from pygmy to Nilotic Negro, also finds its closest counterpart in races of domesticated species. The odd variations in human hair, in length, texture, color and the like, compare with the coat variations of domestic animals—as do man's facial variations. I suppose one could attribute man's tendency to hairlessness either to domestication or to fetalization.

And then there is the matter of human sexual behavior. Domesticated animals tend to lose seasonal sexual periodicity, and the males in particular tend to be able to respond sexually at any time. But this is true also of the undomesticated monkeys and apes. The lack of periodicity in sexual response on the part of the human female is, as far as I know, unique, and I don't know how to explain it in biological terms. The whole human preoccupation with sex looks like a cultural development: only an animal with an adequate and assured food supply, free from the constant hazards of the wild, could afford to separate sexual behavior so completely from reproductive needs. Yet the continuing sexuality of the human adult is undoubtedly a strong factor in the social cohesiveness of human groups—in other words, it looks as though man's sexual proclivities helped make culture possible, while culture, in turn, enabled man to devote an unusual amount of energy to sex.

The discussion of domestication or self-domestication can thus hardly be divorced from the discussion of human culture —which is a large subject. But whether we look at man himself, whether we look at the whole complex human association, or whether we look at some single aspect of the association, we find that we are dealing with a curious and in many ways exceptional aspect of the biosphere. We come up against the fact that man can be explained as a part of the natural system, but that in many ways, he acts as though he were apart from nature—which poses the question of man's place in nature.

16. Man's Place in Nature

We cannot command nature except by obeying her.

—FRANCIS BACON, in *Novum Organum*

We started this book in the tropical forest, thinking about the sea, looking at the similarities in the way life is organized in these different circumstances. This led us to some reflections on the continuity of life in space, in the biosphere, and on its continuity in time, in evolution. The grand design of this system of life includes many different patterns—seas, reefs, lakes, rivers, forests, grasslands and deserts—and we looked at some of these. The structure of the system is everywhere similar, though everywhere complex, turning on the relations among individuals, populations and communities of organisms, and on relations with the physical environment. We have only glanced at these relations, though I hope the glance has been lingering enough to reveal some of the infinite possibilities for study and contemplation.

Then we came to man and his place in this system of life. We could have left man out, playing the ecological game of "let's pretend man doesn't exist." But this seems as unfair as the corresponding game of the economists, "let's pretend nature doesn't exist." The economy of nature and the ecology of man are inseparable and attempts to separate them are more than misleading, they are dangerous. Man's destiny is tied to nature's destiny and the arrogance of the engineering mind does not change this. Man may be a very peculiar animal, but he is still a part of the system of nature.

From the point of view of zoological classification, man is easy enough to deal with. To be sure, there are all sorts of variations among men in color of skin, eyes and hair; in texture of hair and in its distribution over the body; in details of the face and in bodily physique. But no special study is needed to show that these varieties breed together easily and freely whenever they come in contact: that they form a single species.

The classification of the varieties does present problems. They can be lumped together into three or four main types, or they can be split into thirty or forty different races. Whatever system is used, many individuals will be found that do not fit into any category. The varieties, however, show a rough geographical pattern and it is likely that the differences arose through geographical separation, a common enough phenomenon with many kinds of animals.

This human species, as I pointed out, can logically be placed in a genus and family by itself; and men, along with the great apes, Old World monkeys, New World monkeys and a few other animals, can be grouped as an order (the primates) in the general class of mammals.

From the point of view of ecology, man is less easy to deal with. Essentially he is a predatory animal, a second- or third-order consumer. But he shows a tremendous variety of food habits and in many parts of the world he is primarily a first-order consumer, living directly off vegetation. This is probably a rather recent development (in geological terms) because man's plant-eating habits depend to a large degree on processing with fire. The tubers and grains that make up the basic starches of the human diet in most parts of the world are inedible for man unless cooked. He can and does, on the other hand, eat meat of all sorts (including fish and molluscs) without cooking. His vegetable diet, without fire, would be limited to things like fruits and nuts.

It is hard to define the human habitat, because men are found everywhere on land and around the margins of the seas, except under conditions of extreme dryness or extreme cold. But this wide ecological distribution depends on cultural rather than biological traits: on the use of fire, the wearing of clothes, the construction of shelters, the management of boats. It looks as though naked, uncultured man would be a tropical or subtropical species adapted to the rain forest or to transition zones between forest and scrub or forest and grassland. But we have no specimens of uncultured man for study.

When we try to study the relations of man with the physical and biological environment, we always come up against the problem of how to deal with cultural traits. Should we consider culture as a part of man, as an essential attribute of the human species; or should we consider culture as a part of the environment in which this human species lives? This seems like a quibbling, academic sort of question, but it has worried me for a long time. And this specific and striking case has contributed to my general feeling that it is often misleading to attempt to distinguish sharply between organism and environment, whether dealing with men or mice. We are always concerned with interacting systems—which sometimes act as single systems.

I have no substitute for the idea of environment, and I wouldn't know how to get along without the word, but it is tricky. To get back to the problem of man: it seems to me that in general, psychologists tend to treat culture as a part of the environment, while anthropologists tend to regard culture as a part of man, as a part of his equipment for dealing with the environment. To put it another way, psychologists tend to regard culture as a constant, something to which all men are subjected, and they are interested in the ways in which individuals cope with this: how they learn to conform or, consciously or unconsciously, to rebel. They find that frustrations, joys, neuroses, psychoses, all sorts of human behavioral patterns, derive from the interactions between animal man and this cultural environment. At least it often seems to me that this is what they are doing—though the psychologists themselves are not particularly apt to use the world culture.

In contrast, anthropologists tend to take "human nature" for granted, to regard animal man as more or less the same everywhere and to explain all differences in cultural terms. They find culture to be adaptive: the Eskimo way of life fitted to the arctic and subarctic environment, the Dyak way of life to the rain forest conditions of Borneo. They are preoccupied with the description and analysis of all the different kinds of culture they can find, with the study of cultural evolution, and with the effects of cultural diffusion and contact.

In its extreme form, this point of view in anthropology ignores animal man altogether. Culture, I suppose they would have to admit, could not continue without continuing men, and men in this sense created the cultural forces. But men have long since become the helpless victims of their cultures, and developments go on inexorably, according to the laws governing cultural evolution, regardless of the will or desire or power of any individual.

From the point of view of biology, it is most convenient to treat culture sometimes as an attribute of man, sometimes as a part of the human environment. The biologist, trying to look at the human species, cannot think in terms of man and environment: he must deal always with man-culture-environment. Pygmies in the Congo, Bantu in the Congo, Belgians in the Congo, are all men, *Homo sapiens,* in a particular geographical and ecological environment. But they behave quite differently; the environment has quite different meanings for them; they, in turn, have quite different effects on the environment. The whole ecological situation is different.

The problem of man's place in nature, then, is the problem of the relations between man's developing cultures and other aspects of the biosphere. The understanding of these is greatly handicapped by the way in which we have come to organize knowledge. To be sure, man with his varying cultures and cultural traits forms a special phenomenon which requires special means of study and the accumulation of special sorts of information. But still, man has not escaped from the biosphere. He has got into a new, unprecedented kind of relationship with the biosphere; and his success in maintaining this may well depend not only on his understanding of himself, but on his understanding of this world in which he lives.

This makes the split between the social and biological sciences particularly unfortunate. Economics and ecology, as words, have the same root; but that is about all they have in common. As fields of knowledge, they are cultivated in remotely separated parts of our universities, through the use of quite different methods, by scholars who would hardly recognize anything in common. The world of the ecologists is "unspoiled nature." They tend to avoid cities, parks, fields, orchards. The real world of the economists is like Plato's, it is a world of ideas, of abstractions—money, labor, market, goods, capital. There is no room for squirrels scolding in the oak trees, no room for robins on the lawn. There is no room for people either, for that matter—people loving and hating and dreaming. People become the labor force or the market.

More and more, in all areas, we tend to separate the study of man from the study of nature. The separation is one of the basic lines of division in the way we have organized knowledge, in our pattern of specialization. The natural sciences and the social sciences exist in practically complete isolation from one another. Man's body, curiously, has been left with the natural sciences while the social sciences have taken over his mind—at a time when we are most aware of the artificiality of the body-mind separation.

Our third great division of basic knowledge, the humanities, has long since forgotten about nature. Joseph Wood Krutch can well remark: "There are many courses in 'The Nature

Poets' in American colleges. But nature is usually left out of them." Surely there is some way of putting all of these things together, of achieving a more balanced view of ourselves and the rest of the natural world. The matter, I think, has some urgency.

Ours has been aptly called the age of anxiety, and this is curious. We should be able to look about us and feel a certain self-satisfaction. We have learned to develop and direct tremendous power; we can create the kind of conditions we find comfortable; we can produce large quantities of a great variety of foods; we have achieved a surprising degree of control over disease and physical pain. In almost any way we assess man's relations with his environment, he seems to be doing well when compared with the past, even though there is still obvious room for great improvement.

Yet, despite this abundance and progress, almost all attempts to look at man's future are gloomy. I can't think of any recently written image of the future that sounds very attractive, even when the author was trying hard to look for glories. The glories mostly turn out to be bigger and better gadgets, faster trips to a dismal Mars, or better adjusted husbands and wives who no longer take to drink. Usually the author looking into the future doesn't pretend to like his 1984 or his brave new world: but looking about him, this is what he sees coming.

Our anxiety about the future, when we analyze it, turns largely on three related things: the likelihood of continuing warfare, the dizzy rate of human population growth, and the exhaustion of resources. But these don't look like insoluble problems. Surely men who can manufacture a moon can learn to stop killing each other; men who can control infectious disease can learn to breed more thoughtfully than guinea pigs; men who can measure the universe can learn to act wisely in handling the materials of the universe. Why are we so pessimistic?

Chiefly, I suspect, because we have come more and more to doubt our ability to act rationally. Reason seems to be a property of individual men, not of the species or of organized groups. Somewhere we have lost the faith of the Eighteenth Century French philosophers in the perfectibility of man, and the rather different faith of the Nineteenth Century in the idea of progress.

Maybe the anthropologists are right when they say that culture acts as a thing in itself, sweeping along according to inexorable laws, no more under man's control than rodent evolution is under the control of the mice in the fields. The difference between men and mice, then, would be a matter only of awareness, of self-consciousness. We can study the laws of cultural evolution—or organic evolution—but we can't change them. We can foretell our doom but we can't forestall it.

I don't believe this, and I doubt whether the extreme culturists really believe it either. If they believed what they say, I think they wouldn't talk so much. They are like the disciples of Karl Marx who say they believe in the inexorable dialectic of history, but continually try to give history a push in the right direction. Man can't change the laws of cultural evolution or organic evolution—true enough, no doubt—but understanding the laws and acting with the laws, he can influence the consequences. He has in his hands a certain measure of control over his destiny, but this control depends on understanding, and on the spread and proper use of knowledge.

The great immediate threat, of course, is the misuse of nuclear power, the danger of catastrophic war. The long-term threat is the cancerous multiplication of the numbers of men: a new human population the size of the city of Detroit every month, year after year. The thought is dizzying. And then the thought of a nuclear blast capable of killing last month's millions in a few seconds is hardly reassuring. It looks as though, as a part of nature, we have become a disease of nature—perhaps a fatal disease. And when the host dies, so does the pathogen.

How, in the face of our power, in the face of our danger, do we develop a guiding philosophy?

No single man, no single field of knowledge, holds the answer to that. But all men and all knowledge can contribute to the answer. Insofar as man's relations with the rest of nature are concerned, I think we must make every effort to maintain diversity—that we must make this effort even though it requires constant compromise with apparent immediate needs. To look at this, it may be most convenient to sort out the arguments into those that are primarily ethical, those that are primarily esthetic, and those that are essentially utilitarian.

Albert Schweitzer remarks in his autobiography that "the great fault of all ethics hitherto has been that they believed themselves to have to deal only with the relations of man to man." This is particularly true in the Western, Christian tradition. The present material world, in the philosophy of this tradition, is unimportant, no more than a transient scene for the testing of the soul's fitness for eternity. The material universe is completely man-centered. Nature, insofar as it is noticed, is only a convenience—or a temptation—with no positive value in itself.

Animals are unimportant because they have no souls. God may notice the sparrows, but this is an example of His omniscience rather than of His preoccupation. Even Christ gave no thought to the Gadarene swine. The first arguments against bear-baiting, cockfighting and the like were not that they were liable to cause injury and pain to the animals, but that they were liable to demoralize the human character, leading to gambling, thievery and the like.

For a considerable part of humanity, however, this world has direct religious significance. Many primitive religions have various forms of nature worship, of animism and totemism. But in some of the great religions, particularly Buddhism and Hinduism, attitudes toward nature—toward animals in particular—have an ethical basis. For many millions of Hindus it is a sin to kill any animal. With the Jains, this is carried to an extreme to avoid possible injury even to the tiniest of insects.

We deplore the Hindu attitude toward cattle as uneconomical—which it certainly is—and a handicap to the development of India. In countries within the Western tradition, however, attitudes toward animals often cannot be explained on practical or rational grounds. I suspect that a visitor from Mars, observing our treatment of dogs, cats and other domestic pets, would conclude that they were sacred animals. Horses in some Western subcultures are also treated as sacred animals. The horror of eating horse meat—or dog meat—seems not too different from the Muslim horror of eating pig or the Hindu horror of eating any animal.

There have always been individuals within the Christian tradition with a love of nature, with a kind feeling toward animals. St. Francis of Assisi rightfully is their patron. In modern times this has grown into a cult of great emotional force, leading to the development of a variety of formal organizations for the prevention of cruelty to animals, for the protection of wildlife, which reaches an extreme in the anti-vivisectionist groups. This attitude is most highly developed in the industrialized regions since it goes along with economic security and relative leisure. It is a characteristic of "affluent societies." It is reassuring in the sense that kindness and tolerance and sympathy—whether for slaves, for children or for animals—seem to gain force and spread with economic development.

This kindness and sympathy for animals might well be classed as an ethical attitude. Curiously, along with the cult of kindness to animals, we have a parallel development in the same societies and circumstances of the cult of the sportsman, in which killing becomes a good in itself. As hunting ceased to be a necessity, it became a luxury for men; and hunting as play, hunting as sport, has long characterized classes of men with the leisure to indulge in it. Hunting is sometimes thought to represent a basic "instinct" in human nature, and certainly there is something elemental and primitive in the thrill of the chase. Intellectually, I have abandoned hunting as a sport since, when a boy, I watched the agonies of a raccoon I had wounded. But often enough, hunting for some worthy "scientific" purpose, I have felt my intellectual pretensions slide away and I have become lost in the purely emotional absorption of getting my game.

The sport of kings and noblemen has now become the sport of millions, of anyone with an automobile and a rifle or a shotgun. It is recreation. But also a philosophy has developed whereby this killing of deer and ducks and quail is supposed to inculcate virtue. Krutch quotes the propaganda slogan of a gun company: "Go hunting with your boy and you'll never have to go hunting for him."

I get lost in the ethical issues involved in these problems. Intellectually I sympathize with the teachings of Buddha, that all life is sacred. But practically, I see no way of acting on this. There is no logical stopping place before the end reached by the people of Samuel Butler's *Erewhon*. They became vegetarian out of respect for the rights of animals. But as one of their learned men pointed out, vegetables are equally alive, and equally have rights. So the Erewhonians, to be consistent, are reduced to eating cabbages certified to have died a natural death. Monkeys, deer, cows, rats, quail, songbirds, lizards, fish, insects, molluscs, vegetables—where do you draw the line between what can be properly killed and eaten, and what not? It so happens that I don't like decayed cabbages and I do like rare roast beef—which leaves me, as usual, blundering around in a quandary.

The ethical question is difficult. We have drifted in the modern world into a position of ethical relativism which leaves us with no absolutes of good and bad, right and wrong. Things are good or right according to the context, depending on the values of the society or culture. Yet one feels that there must be some basis of right conduct, applicable to all men and all places and not depending on any particular dogma or any specific revelation. Science has undermined the dogmas and revelations; and it provides, for many working scientists, a sort of faith, a sort of humanism, that can replace the need for an articulated code of conduct. But our scientists and philosophers have so far failed to explain this in a way that reaches any very large number of people. This, it seems to me, is one of the great tasks of modern philosophy, which the philosophers, dallying in their academic groves, have shunned.

When some thinker does come forth to provide us with a rationale for conduct, he will have to consider not only the problems of man's conduct with his fellow men, but also of man's conduct toward nature. Life is a unity; the biosphere is a complex network of interrelations among all the host of living things. Man, in gaining the godlike quality of awareness, has also acquired a godlike responsibility. The questions of the nature of his relationships with the birds and the beasts, with the trees of the forests and the fish of the seas, become ethical questions: questions of what is good and right not only for man himself, but for the living world as a whole. In the words of Aldo Leopold, we need to develop an ecological conscience.

It is sometimes said that the esthetic appreciation of nature is relatively new, that the Greeks, for instance, did not admire landscapes. The matter can be argued and I don't know that anyone has made a careful study of changing attitudes, or of differences in attitude among the great civilizations. Within our own civilization, it looks as though the conscious appreciation of the beauties of nature had its roots in the so-called Romantic Movement of the Eighteenth Century. We can see this most plainly in literature, in landscape painting and in landscape architecture. It is less clear in the other arts, though Lovejoy plausibly equates it with the love of diversity and the search for new forms that characterize Western art generally in the last two centuries.

It looks as though man's esthetic appreciation of nature increases as the development of his civilization removes him from constant and immediate contact with nature. The peasant hardly notices the grandeur of the view from his fields; the woodsman is not impressed by stately trees, nor the fisherman by the forms and colors on the reefs. In part, this is the general problem of not seeing the familiar, of not appreciating what we have until it is lost.

The reasons behind the conservation movement, from this point of view, are similar to the reasons for preserving antiquities, for maintaining museums of art or history or science. Nature is beautiful, therefore it should not be wantonly destroyed. Representative landscapes should be preserved because of their esthetic value, because of their importance in scientific study, and because of their possibilities for recreation.

I have often wished, as I saw a tropical forest being cleared, that this beautiful place could somehow be protected and preserved for the future to enjoy. The idea, to the people involved in the clearing, seems absurd. The forest is an enemy, to be fought and destroyed; beauty lies in the fields and orchards that will replace it. This was the attitude of our ancestors who in the end effectively cleared the great deciduous forest that once covered the eastern United States, leaving only accidental and incidental traces. How we would love now to have a fair sample of that great forest! But the idea of deliberately saving a part of the wilderness they were conquering never occurred to the pioneers. Nor does it occur to pioneers now in parts of the world where pioneering is still possible.

There must be some way in which one nation can profit by the experience of another nation; some way of saving examples of the landscapes and wildlife that have not yet been devastated by the onrush of industrial civilization. In Africa there is a danger that the national parks will be regarded as toys of colonial administrations, and fade with the fading of those administrations. And the colonial powers, even with the experience of loss in their homelands, are not always too careful about the preservation and maintenance of samples of the natural world under their care.

In tropical America we have the effect of the Spanish tradition. The Romantic Movement never crossed the Pyrenees. Spanish thought and art remain essentially man-centered. Some of my Spanish friends have suggested that the relative failure of science to develop in that tradition may be a consequence of this indifference, on the part of most of the people, to the world of nature. The correction for this might be deliberate attempts to foster nature study in the school systems. Whatever the cause, the conservation movement has not made great headway in the parts of the world dominated by Spanish culture.

In the United States, we have a National Park system, and various sorts of reservations and wildlife refuges under national, state and private auspices. This is largely the consequence of the dedicated efforts of a few people, and we are still far from the point where we can sit back and congratulate ourselves. Conservation interests fall under different branches of government and efforts to form a coherent and unified national policy have not been very successful; we still have no Department of Conservation with cabinet rank. The struggle for financial support is always hard. And there is a constant, eroding pressure from conflicting private and governmental interests.

Ugliness—by any esthetic standard—remains the predominant characteristic of development, of urbanization, of industrialization. We talk about regional planning, diversification, working with the landscape—and we build vast stretches of the new suburbia. The ideas so forcefully developed by Patrick Geddes, Lewis Mumford and others like them, fall on deaf ears. We need an ecological conscience. We also need to develop ecological appreciation. The Romantic Movement, despite its two hundred year history, has not yet reached our city councils or our highway engineers.

Practical considerations are—and perhaps ought to be—overwhelmingly important in governing man's relations with the rest of nature. Utility, at first thought, requires man to concentrate selfishly and arrogantly on his own immediate needs and convenience, to regard nature purely as a subject for exploitation. A little further thought, however, shows the fallacy of this. The danger of complete man-centeredness in relation to nature is like the danger of immediate and thoughtless selfishness everywhere: the momentary gain results in ultimate loss and defeat. "Enlightened self-interest" requires some consideration for the other fellow, for the other nation, for the other point of view; some giving with the taking. This applies with particular force to relations between man and the rest of nature.

The trend of human modification of the biological community is toward simplification. The object of agriculture is to grow pure stands of crops, single species of plants that can be eaten directly by man; or single crops that provide food for animals that can be eaten. The shorter the food chain, the more efficient the conversion of solar energy into human food. The logical end result of this process, sometimes foreseen by science fiction writers, would be the removal of all competing forms of life—with the planet left inhabited by man alone, growing his food in the form of algal soup cultivated in vast tanks. Perhaps ultimately the algae could be dispensed with, and there would be only man, living through chemical manipulations.

Efficient, perhaps; dismal, certainly; and also dangerous. A general principle is gradually emerging from ecological study to the effect that the more complex the biological community, the more stable. The intricate checks and balances among the different populations in a forest or a sea look inefficient and hampering from the point of view of any particular population, but they insure the stability and continuity of the system as a whole and thus, however indirectly, contribute to the survival of particular populations.

Just as health in a nation is, in the long run, promoted by a diversified economy, so is the health of the biosphere promoted by a diversified ecology. The single crop system is always in precarious equilibrium. It is created by man and it has to be maintained by man, ever alert with chemicals and machinery, with no other protection against the hazards of some new development in the wounded natural system. It is man working against nature: an artificial system with the uncertainties of artifacts. Epidemic catastrophe becomes an ever present threat.

This is one of the dangers inherent in man's mad spree of population growth—he is being forced into an ever more arbitrary, more artificial, more precarious relation with the resources of the planet. The other great danger is related. With teeming numbers, an ever tighter system of control becomes necessary. Complex organization, totalitarian government, becomes inevitable; the individual man becomes a worker ant, a sterile robot. This surely is not our inevitable destiny.

I am not advocating a return to the neolithic. Obviously we have to have the most efficient systems possible for agriculture and resource use. But long run efficiency would seem to require certain compromises with nature—hedgerows and woodlots along with orchards and fields, the development of a variegated landscape, leaving some leeway for the checks and balances and diversity of the system of nature.

Ethical, esthetic and utilitarian reasons thus all support the attempt to conserve the diversity of nature. It is morally the right thing to do; it will provide, for future generations, a richer and more satisfying experience than would otherwise be possible; and it provides a much needed insurance against ecological catastrophe. "Unless one merely thinks man was intended to be an all-conquering and sterilizing power in the world," Charles Elton has remarked, "there must be some general basis for understanding what it is best to do. This means looking for some wise principle of co-existence between man and nature, even if it be a modified kind of man and a modified kind of nature. This is what I understand by *conservation*."

In defying nature, in destroying nature, in building an arrogantly selfish, man-centered, artificial world, I do not see how man can gain peace or freedom or joy. I have faith in man's future, faith in the possibilities latent in the human experiment: but it is faith in man as a part of nature, working with the forces that govern the forests and the seas; faith in man sharing life, not destroying it.

Notes and Sources

The ideas in this book have accumulated over many years and often I can no longer remember where I picked them up. Sometimes I thought that a particular idea was original, but I turned out to be mistaken in every case. Friends will no doubt find many of their thoughts appropriated here; that, I suppose, is an inevitable hazard of friendship with an author. I am particularly conscious of the things I have learned from daily conversation with my colleagues in Ann Arbor, especially John Bardach, Fred Smith, Nelson Hairston and Larry Slobodkin. I have tried to observe the faint boundary between research and plagiarism but it is often hard to be sure which side of the boundary you are on.

The material in the book is essentially that of my lectures during the past two years in my undergraduate general education course, "Zoology in Human Affairs." I suppose I'll have to think up a new set of lectures now. A preliminary draft of the manuscript was tried out on a seminar group in "Social Biology" and my debt to all of these students is really enormous.

My title looks like a direct steal from that delightful book by H. M. Tomlinson, *The Sea and the Jungle* (New York: E. P. Dutton & Co., 1920); but I think the similarity is coincidence. I first used the title for an article published in *The Scientific Monthly* (May, 1945) which covered much the same ground as Chapter II of the present book.

G. G. Simpson, C. S. Pittendrigh and L. H. Tiffany have written a thorough and well-organized textbook on biology entitled *Life: An Introduction to College Biology* (New York: Harcourt, Brace, 1957) which has been on my desk at all times—and frequently consulted. I have also frequently used that large and useful compendium by W. C. Allee, A. E. Emerson, Orlando Park, Thomas Park and K. P. Schmidt called *Principles of Animal Ecology* (Philadelphia: Saunders, 1949): a book affectionately known to biology graduate students, from the initials of its distinguished authors, as "The Great Aepps."

Essentially I have been trying to write an introduction to ecology, and there are a number of good ecological textbooks which I have consulted often. By far the best reading is Charles Elton, *Animal Ecology* (New York: Macmillan, 1927 and 1936). Others that I have used are E. P. Odum, *Fundamentals of Ecology* (Philadelphia: Saunders, 1953); G. L. Clarke, *Elements of Ecology* (New York: Wiley, 1954); and Richard Hesse, W. C. Allee and K. P. Schmidt, *Ecological Animal Geography* (New York: Wiley, 1951).

1. The Study of Life

My chapter quote is from a little book by Jean Rostand published in New York by Basic Books in 1959. Otherwise I don't see anything in the chapter to document—it is mostly personal opinion.

2. Landscapes—and Seascapes

My wife wrote what seems to me a delightful account of our years in Villavicencio: Nancy Bell Bates, *East of the Andes and West of Nowhere* (New York: Scribner's, 1947). Anyone who wants to check on the details of local geography might look up my article on "Climate and Vegetation in the Villavicencio Region of Eastern Colombia" in the *Geographical Review,* vol. 38, (1948) 555-574. I told the yellow fever story in an article entitled "The Natural History of Yellow Fever in Colombia" published in *The Scientfic Monthly,* vol. 63, (1946) 42-52.

3. The Living World

Our current understanding of the origin of life has been discussed in an article with that title by George Wald, published in the *Scientific American* (August, 1954) and reprinted in a book by the editors of the *Scientific American, The Physics and Chemistry of Life* (New York: Simon and Schuster, 1955). The effect of man on the biosphere has been very thoroughly covered in a big book edited by W. L. Thomas, Jr., *International Symposium on Man's Role in Changing the Face of the Earth* (University of Chicago Press, 1956). I have consulted articles in this book frequently in the course of my writing.

4. The Open Sea

There is a large literature on the biology of the sea. The standard technical introduction to the subject is H. U. Sverdrup, M. W. Johnson and R. H. Fleming, *The Oceans: Their Physics, Chemistry and General Biology* (New York: Prentice-Hall, 1942). The natural history of plankton has been covered in a beautiful volume in the British New Naturalist Series: Alister C. Hardy, *The Open Sea* (London: Collins, 1956). Another fine volume in the same series is *The Sea Shore* (London: Collins, 1949) by C. M. Yonge. For information about ocean depths I have depended on N. B. Marshall, *Aspects of Deep Sea Biology* (New York: Philosophical Library, 1954). There is also much information in a little book by Klaus Guenther and Kurt Deckert, *Creatures of the Deep Sea* (New York: Scribners, 1956). "The Bathyscaph" has been described in an article by R. S. Dietz, R. V. Lewis and A. B. Rechnitzer in the *Scientific American* (April, 1958).

5. The Coral Reef

Obviously a good part of this chapter comes from discussions with my friend John E. Bardach, who has spent a lot of time skin-diving around the reefs of Bermuda and the Caribbean Sea. I probably have an equally large debt to Donald P. Abbott, who taught me something new about invertebrates every day during the period we were together in Micronesia. I recommend his half of the book we wrote together, *Coral Island* (New York: Scribner's, 1958). There are a number of books on skin-diving and fish-watching. A good one is by Carleton Ray and Elgin Ciampi, *The Underwater Guide to Marine Life* (New York: Barnes, 1956).

6. Lakes and Rivers

There are several textbooks on fresh water biology or limnology, but they make rather dull reading. My favorite is an old one by Kathleen E. Carpenter, *Life in Inland Waters* (New York: Macmillan, 1928). Robert E. Coker has written an account of inland waters for the general reader, *Streams, Lakes and Ponds* (Chapel Hill: University of North Carolina Press, 1954). The story of the migrations of eels and salmon has been told by Lorus and Margery Milne in *Paths Across the Earth* (New York: Harper, 1958); and more dramatically and thus perhaps less informatively by Georges Blond, *The Great Migrations* (New York: Macmillan, 1956). My account of the Teays River is based on an article by Raymond E. Janssen, "The History of a River" in the *Scientific American* (June, 1952). My figures on the size of fish come from Leonard P. Schultz and Edith M. Stern, *The Ways of Fishes* (New York: Van Nostrand, 1948). An excellent book on fish in general, from fresh or salt water, is J. R. Norman, *A History of Fishes* (New York: A. A. Wyn, 1947).

7. The Rain Forest

My figures on the maximum height of trees come from P. W. Richards, *The Tropical Rain Forest* (Cambridge University Press, 1952). Many beautiful photographs on rain forest vegetation and animals are included in the book by E. A. de la Rue, François Bourlière and Jean-Paul Harroy, *The Tropics* (New York: Knopf, 1957).

8. Woodland, Savanna and Desert

The general books on ecology all include accounts of tundra, taiga and deciduous forest. The classical quotations do not reflect my own scholarship, they are taken from the chapter by H. C. Darby, "The Clearing of the Woodland in Europe," in the book edited by W. L. Thomas, Jr., *Man's Role in Changing the Face of the Earth.* Many other parts of this book have material relevant to this chapter, with detailed bibliographies. There is an interesting article on "Ecology of Desert Plants" by Fritz Went in the *Scientific American* (April, 1955) and P. A. Buxton has written a book on *Animal Life in Deserts* (London: Arnold, 1923).

9. The Units of Life

An enormous number of articles and books on population biology has appeared in recent years. It is always discussed in the ecology textbooks though I think the best general summary is still that made by Allee and others in *Principles of Animal Ecology.* I tried to deal with the subject in a rather light way in my book on *The Prevalence of People* (New York: Scribners, 1955) where I included a bibliography.

10. The Biological Community

This subject is also discussed in all ecology textbooks, where appropriate bibliographies can be found.

11. The Natural History of Disease

Some of the ideas in this chapter are developed at greater length in a section I wrote on "The Ecology of Health" for a book edited by Iago Galdston, *Medicine and Anthropology* (New York: International Universities Press, 1959). I have got many ideas about the nature of health and of death from Peter B. Medawar, *The Uniqueness of the Individual* (New York: Basic Books, 1958). Of the various textbooks of parasitology, the one I find most interesting is T. W. M. Cameron, *Parasites and Parasitism* (New York: Wiley, 1956).

12. Animal Behavior

General reviews of our knowledge of animal behavior have been written by Nikolaas Tinbergen, *The Study of Instinct* (Oxford: Clarendon Press, 1951); W. H. Thorpe, *Learning and Instinct in Animals* (London: Methuen, 1956) and John P. Scott, *Animal Behavior* (University of Chicago Press, 1958). Of these, I would recommend looking at Thorpe first. There is a great deal of information, with extensive bibliographies, in a book edited by Anne Roe and G. G. Simpson, *Behavior and Evolution* (New Haven: Yale University Press, 1958). Anyone who likes animals and wants to gain insights into their behavior will enjoy reading the book by Konrad Lorenz, *King Solomon's Ring* (New York: Crowell, 1952). I think the whole problem of understanding behavior in terms of the perceptual environment of the animal concerned is best illustrated by the studies of echo-location which have been summarized by Donald R. Griffin, *Listening in the Dark* (New Haven: Yale University Press, 1958).

13. Social Life Among the Animals

There are many books on the social insects. A good introduction is by Charles D. and Mary Michener, *American Social Insects* (New York: Van Nostrand, 1951). An older book that still reads well is W. M. Wheeler, *Social Life among the Insects* (New York: Harcourt, Brace, 1923). Communication in honeybees has been described by Karl von Frisch, *The Dancing Bees* (New York: Harcourt, Brace, 1955). Vertebrate social behavior has been reviewed by Nikolaas Tinbergen, *Social Behavior in Animals* (New York: Wiley, 1953). The best introduction to the behavior of monkeys and apes is still the book by Earnest Hooton, *Man's Poor Relations* (New York: Doubleday, Doran, 1942). C. R. Carpenter's classic studies of primate behavior are "A Field Study of the Behavior and Social Relations of Howling Monkeys," *Comparative Psychological Monographs,* vol. 10, no. 2, 1934; and "A Field Study in Siam of the Behavior and Social Relations of the Gibbon" in the same series, vol. 16, no. 5, 1940. Robert M. Yerkes has written the standard book, *Chimpanzees* (New Haven: Yale University Press, 1943). Most everything we know about gorillas in the wild is in a book by Fred G. Merfield and Harry Miller, *Gorilla Hunter* (New York: Farrar, Straus and Cudahy, 1956).

14. The Human Species

There must be an enormous number of books on human nature and human evolution. I have read quite a few of them and each one has probably had some effect on my thinking. My greatest conscious debt is for ideas gleaned during conversations with S. L. Washburn. He has sketched his theory of human evolution in a chapter written with Virginia Avis on "Evolution of Human Behavior" in the book edited by Anne Roe and G. G. Simpson, *Behavior and Evolution.* The ecological approach, which I have tried to use here, has been reviewed by G. A. Bartholomew and J. B. Birdsell in an article entitled "Ecology and the Protohominids" in the *American Anthropologist,* vol. 55, (1953) 481-498.

A book that I read in proof after this manuscript was completed has impressed me as an excellent introduction for the general reader to contemporary research on human evolution. It is *Adventures with the Missing Link* by Raymond A. Dart and Dennis Craig (New York: Harper, 1959). The book is mostly about Dart's own experiences with the australopithecine fossils, but it also covers a great deal of other material easily and interestingly. For a biologist's approach to the evolution of the human mind, I recommend N. J. Berrill, *Man's Emerging Mind* (New York: Dodd, Mead, 1955). *The Human Animal* by Weston LaBarre (University of Chicago Press, 1954) shows the relation of psychoanalytic theory to this problem. *The Story of Man* by Carleton S. Coon is a well-written and well-rounded anthropological treatment.

15. On Being Domesticated

My ideas about the domestication of animals and plants come—probably obviously enough—from my friendship with two extraordinary people, Carl Sauer and Edgar Anderson. For an introduction to Sauer, I suggest *Agricultural Origins and Dispersals* (New York: American Geographical Society, 1952); to Anderson, *Plants, Man and Life* (Boston: Little, Brown, 1952). The bibliographies of these books provide a key to the general literature on the subject.

16. Man's Place in Nature

I must confess that I got my chapter quote, not directly from Bacon, but from Roderick Seidenberg, *Posthistoric Man* (Chapel Hill: University of North Carolina Press, 1950; reprinted by Beacon Press, 1957). I don't think I got much else from this much-discussed book: but I suppose each man has his own vision (or nightmare) of the human future. The book by Harrison Brown, *The Challenge of Man's Future* (New York: Viking, 1954) seems to me the most balanced appraisal of our problems. The view of culture as a thing in itself is persuasively presented by Leslie White in *The Science of Culture* (New York: Farrar, Straus, 1949). My quotations in this chapter from Joseph Wood Krutch are from *The Great Chain of Life* (Boston: Houghton, Mifflin, 1957) though other books by Krutch have probably had a greater influence on my general thinking.

I had a great deal of trouble in writing this chapter because I kept getting off on the various attempts to derive ethical systems from science or nature, though my topic was the application of ethics to man's relations with nature. The basis of ethics is a fascinating subject, but for a quite different book. I cannot refrain, however, from referring the reader to the last chapters of G. G. Simpson, *The Meaning of Evolution* (New Haven: Yale University Press, 1949; reprinted as a Mentor Book, 1951) for a discussion of the varied attempts to base ethical systems on evolutionary theories.

While writing the section on ethics, I consulted Edward Westermarck, *Christianity and Morals* (London: Kegan Paul, 1939) which has a chapter on "Christianity and the Regard for Lower Animals." The idea of an ecological conscience comes from Aldo Leopold, *A Sand County Almanac* (New York: Oxford University Press, 1949). My debt at various points to Arthur Lovejoy, *The Great Chain of Being* (Cambridge: Harvard University Press, 1936) is probably obvious enough. But the whole idea of trying to look at man's relations with nature in terms of ethics, esthetics and utility comes from Charles Elton, *The Ecology of Invasions by Animals and Plants* (New York: Wiley, 1958).

Index

A Note about the Production of This Book

The typeface for the text of this special edition of *The Forest and the Sea* is Times Roman, designed in 1932 by Stanley Morison for the London *Times* and first used by that newspaper. The type was set by The Haddon Craftsmen, Inc., of Scranton, Pennsylvania, and printed by The Safran Printing Company of Detroit, Michigan. The binding was done by J. W. Clement Co. of Buffalo, New York. The cover was printed by Livermore and Knight Co., a division of Printing Corporation of America, in Providence, Rhode Island.

⧗

The paper, TIME Reading Text, is from The Mead Corporation of Dayton, Ohio. The cover stock is from The Plastic Coating Corporation of Holyoke, Massachusetts.